Interfaces in Heterogeneous Ceramic Systems

Interfaces in Heterogeneous Ceramic Systems

Ceramics Transactions Series, Volume 191

Proceedings of the 6th Pacific Rim Conference on Ceramic and Glass Technology (PacRim6); September 11–16, 2005; Maui, Hawaii

Edited by
L. P. Cook
S.-i. Tanaka
W. Wong-Ng
R. Schwartz

The American Ceramic Society

WILEY-INTERSCIENCE
A JOHN WILEY & SONS, INC., PUBLICATION

Published by John Wiley & Sons, Inc., Hoboken, New Jersey
Published simultaneously in Canada.

Library of Congress Cataloging-in-Publication Data is available.

ISBN-13 978-0-470-08388-8
ISBN-10 0-470-08388-3

10 9 8 7 6 5 4 3 2 1

Contents

Preface

Most materials have internal interfaces by necessity, or by design, and of course all materials have at least one interface defining their physical extent. Interfacial behavior is in general very important in determining the overall properties of a material, and its suitability for a particular application. Aside from grain boundaries in polycrystalline single-phase materials, the majority of such interfaces are heterogeneous, that is they involve different phases in contact. Consequently, a variety of phenomena become important if the material is placed in a reactive environment, or if it is heated, including crystal growth, diffusion, solution, and surface tension effects. The evolution of the interface during thermal processing, or during service, should be well-understood to achieve the desired result.

This proceedings volume contains papers which were presented as part of the symposium on "Thermochemistry and Metrology of Interfaces," at the 6th Pacific Rim Conference on Ceramic and Glass Technology held September 12 -16, 2005, in Maui, HI. The aim of the symposium was to bring together experts from Pacific Rim countries to discuss research on ceramic-based interfaces in a variety of fields. The intent was to provide a forum for cutting across conventional disciplinary divides, and to focus on the phenomenology of interfacial behavior in a variety of materials having to do with the general field of ceramics. The book is organized in five parts. The first part contains relevant papers on theory and modeling, and includes both first-principles- as well as more empirically-based approaches for predicting and describing interfacial evolution. The second part focuses on wetting phenomena, both of oxides by metal liquids and of metals by oxide liquids, phenomena of obvious importance to the steel industry, among others. The third section focuses on heterogeneous interfaces in high-temperature superconductors, complex materials for which a wealth of data has been accumulating over the past few years. Bio-interfaces, an area of increasing interest as the field of biomaterials continues to advance, are represented in the fourth section. New developments in instrumentation have continuously led to improved characterization of interfaces, as examples described in the fifth section illustrate.

The editors would like to acknowledge the assistance of The American Ceramic

Society staff who have provided essential encouragement and support in preparing this volume. In addition, we thank all of the presenters, authors, session chairs, manuscript reviewers, and ACerS program coordinators and officials who made the symposium and the proceedings possible.

LAWRENCE COOK
SHUN-ICHIRO TANAKA
WINNIE WONG-NG
ROBERT SCHWARTZ

Theory and Modeling

HYBRID QUANTUM-CLASSICAL SIMULATION STUDY OF NANO-INTERFACES

Shuji Ogata, Takahisa Kouno
Graduate School of Engineering, Nagoya Institute of Technology
Gokiso-cho, Showa-ku, Nagoya, Aichi 466-8555, Japan

Takahiro Igarashi
Energy System Department, Japan Atomic Energy Research Institute
Tokai, Naka, Ibaraki 319-1195, Japan

ABSTRACT

Feature sizes of useful electronic devices and micro-electron-mechanics-systems are becoming smaller and reaching nanometer ranges. There is increasing demand to perform dynamic, atomistic simulation of nano-interfaces in the nano-devices with realistic sizes using multiscale, hybrid simulation schemes. Interesting processes include oxidation and diffusion at the nano-scales. We have been developing the hybrid simulation schemes for dynamic simulations of such processes on computation grids, by combining electronic-density-functional (DFT), molecular dynamics (MD), and coarse-grained (CG) particle methods. In the hybrid scheme, a quantum (QM) region composed of a relatively small number of atoms and treated with the DFT, is embedded in a large-scale classical (CL) atomistic system composed of the MD atoms interacting through an empirical inter-atomic potential. The CG particle method gives the boundary conditions for the CL system. The CG particle method is a continuum mechanical approach and is well suited to its connection with an atomistic method. For reasonable coupling of the QM and the CL atomistic regions, a novel hybridization scheme is proposed, which requires no link-atoms and is therefore applicable to a wide range of materials and settings. During the hybrid QM-CL simulation run, size and shape of a QM region change adaptively by monitoring inter-particle distances. The hybrid QM-CL simulation scheme is successfully applied to adsorption and dissociation of H_2O molecules on stressed Si-surfaces and stress-dependent diffusion of an O atom in Si crystals. In both simulations, significant effects of the applied stress on the processes are observed.

INTRODUCTION

Though the MD simulation method has played important roles in computational materials researches,[1-3] it contains two fundamental limitations: One is that empirical inter-atomic potentials used in the MD simulations often fail to describe chemical reactions in materials, such as oxidation and fracture. Another is that direct, atomistic treatment of a material with realistic size is difficult due to its huge number of atoms. Simulation with the realistic size is often necessary since heat produced in the reacting region should be correctly transferred to surrounding regions. Considering these, we have been working on the project of developing hybrid simulation schemes,[4-8] in which the electronic-structure calculation such as the DFT is embedded in the MD method and the MD region connects to the peripheral, continuum region. In this paper we explain recent developments in the project.

Computationally efficient QM approaches based on the DFT have advanced greatly.[9] Recently hybrid DFT-MD simulation schemes have also been developed using the so-called link-atom method[4-7] to couple between the DFT atoms and the MD atoms. Successful applications are found, for example, in Ref. 10. However, we find that the hybrid DFT-MD results with the link-atom scheme are sensitive to selection of the QM atoms. To overcome the difficulty, we have proposed a novel hybrid scheme, called the buffered-cluster method,[8] which requires no link-atoms.

The CG particle method[11] is one of the continuum models with many interesting features suitable to its hybridization with the atomistic method. However, original formulation of the CG particle method contains problems that hinder its applications to real systems. We implement the CG particle method with necessary approximations to overcome the problems.

In Methodology section, both the buffered-cluster method in the hybrid QM-CL simulation and the implementation of the CG particle method are explained. In Applications section, the hybrid QM-CL simulation is applied to interesting engineering problems.

METHODOLOGY

Buffered-cluster method[8]

We consider a model system of crystalline Si to explain the buffered-cluster method in the hybrid QM-CL simulation scheme. The system contains mutually non-overlapping regions of reasonable sizes to be treated with a QM method; atoms in those QM regions are called the QM atoms. The rest of the QM atoms are the CL atoms. We assume the empirical, classical inter-atomic potential by Stillinger and Weber[12] (SW) for the CL calculation of Si. We choose the DFT as the QM method by considering balance between the accuracy and the computation speed, though any calculation method that predicts the electronic structures of the atomic cluster may be used.

To describe the total energy of the hybridized system, we follow the modular approach[4,5] and write it in a linear combination form. The Hamiltonian that predicts dynamics of all the atoms is:

$$H(\vec{R}_{\text{all}},\vec{P}_{\text{all}}) = H_{\text{CL}}^{\text{system}}(\vec{R}_{\text{all}},\vec{P}_{\text{all}}) + \sum_{\text{cluster}}(E_{\text{QM}}^{\text{cluster}} - E_{\text{CL}}^{\text{cluster}}), \tag{1}$$

where $\vec{R}_{\text{all}}$ and $\vec{P}_{\text{all}}$ represent sets of positions and momenta of all the atoms, respectively. The $H_{\text{CL}}^{\text{system}}$ in Eq. (1) is the classical Hamiltonian for the total system composed of the kinetic energy and the potential energy terms: $H_{\text{CL}}^{\text{system}} = E_{\text{kin}}(\vec{P}_{\text{all}}) + E_{\text{CL}}^{\text{system}}(\vec{R}_{\text{all}})$. The last two terms on the right hand side of Eq. (1) may be understood as a quantum correction to the classical potential energy for the QM region. We denote sets of positions of the CL and the QM atoms as $\{\vec{r}_{\text{CL}}\}$ and $\{\vec{r}_{\text{QM}}\}$, respectively; that is, $\vec{R}_{\text{all}} = \{\vec{r}_{\text{CL}}\} + \{\vec{r}_{\text{QM}}\}$.

In the buffered-cluster method,[8] the cluster-energy terms on the right hand side of Eq. (1) are functions of positions of the QM atoms $\{\vec{r}_{\text{QM}}\}$ only:

$$E_{\text{QM}}^{\text{cluster}} = E_{\text{QM}}^{\text{cluster}}(\{\vec{r}_{\text{QM}}\}), \tag{2}$$

$$E_{\text{CL}}^{\text{cluster}} = E_{\text{CL}}^{\text{cluster}}(\{\vec{r}_{\text{QM}}\}). \tag{3}$$

In both QM and CL calculations of the QM region in Eqs. (2) and (3), additional atoms are placed at the dangling-bond sites of the QM atoms at the QM-CL boundaries to buffer possible effects of the cutting bonds. We call those atoms as the buffer atoms.

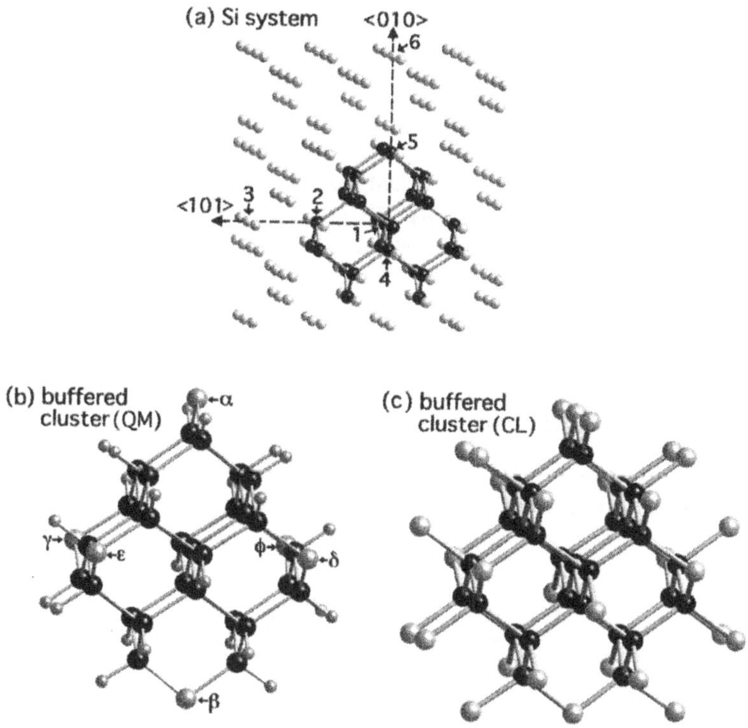

Fig.1: (a) The QM atoms (black spheres) chosen in the hybrid QM-CL simulation of crystalline Si with the buffered-cluster method. Atoms-1, 2, and 3 are equivalent in the crystalline symmetry; likewise atoms-4, 5, and 6 are equivalent. (b) The atomic cluster for the QM calculation of the QM region depicted in (a) with the buffered-cluster method. The large spheres in grey labeled with the Greek letters correspond to the buffer-Si atoms, while the small spheres in grey to the buffer-H atoms. (c) The atomic cluster for the CL calculation of the QM region depicted in (a) with the buffered-cluster method. The white spheres denote the buffer-Si atoms.

For the CL calculation of the QM region in Eq. (3), the buffer atoms are Si to mimic the original bonds at the QM-CL boundary. Starting from positions of either the corresponding CL-Si atoms or the buffer-Si atoms at the last step, positions of the buffer-Si atoms are adjusted at every time step in the simulation run to minimize the potential energy $E_{CL}^{cluster}$ under the constraint of fixing $\{\vec{r}_{QM}\}$. Here, the buffer-Si atoms interact only with the bonding-pair Si-atoms in the original system through the two- and the three-body terms of the SW inter-atomic potential. An example of the QM region in crystalline Si is shown in Fig. 1(a), in which the black spheres denote the QM atoms and the white spheres the CL atoms. Relating to the QM region chosen in Fig. 1(a), the atomic cluster for the CL calculation contains the buffer-Si atoms as depicted in Fig. 1(c). Note that positions of the buffer atoms in the buffered-cluster method may differ from that of the corresponding CL atoms in the total system.

For the QM calculation of the QM region in Eq. (2), the buffer atoms are either H or Si depending on the bonding characteristics in the original system, as shown in Fig. 1(b): the small spheres in grey denote the buffer-H atoms and the large spheres in grey the buffer-Si atoms. Positions of the buffer atoms (either H or Si) for the QM calculation are determined by referring to that of the buffer atoms in the atomic cluster for the CL calculation in Fig. 1(c). If a buffer-Si atom located at $\vec{r}_{\rm CL}^{\rm buffer}$ connects to a single QM atom at $\vec{r}_{\rm QM}$ in the atomic cluster for the CL calculation, we put a buffer-H atom at $\vec{r}_{\rm b} = \beta\vec{r}_{\rm CL}^{\rm buffer} + (1-\beta)\vec{r}_{\rm QM}$ with the scaling factor $\beta = 0.66$ in the atomic cluster for the QM calculation. Otherwise, we put a buffer-Si atom at $\vec{r}_{\rm b} = \vec{r}_{\rm CL}^{\rm buffer}$ in the atomic cluster for the QM calculation. We note that positions of the buffer atoms are not relaxed in the atomic cluster for the QM calculation.

Since both $E_{\rm QM}^{\rm cluster}$ and $E_{\rm CL}^{\rm cluster}$ in Eqs. (2) and (3) are independent of positions of the CL atoms $\{\vec{r}_{\rm CL}\}$, the QM atoms interact with the CL atoms through the classical inter-atomic potential in $H_{\rm CL}^{\rm system}$. Explicit formulas of the atomic forces in the buffered-cluster method are the following. For QM atom-i, the force

$$\vec{F}_i = -\frac{\partial H}{\partial \vec{r}_{{\rm QM},i}} = -\frac{\partial E_{\rm CL}^{\rm system}}{\partial \vec{r}_{{\rm QM},i}} - \sum_{\rm cluster}\left(\frac{\partial E_{\rm QM}^{\rm cluster}}{\partial \vec{r}_{{\rm QM},i}} - \frac{\partial E_{\rm CL}^{\rm cluster}}{\partial \vec{r}_{{\rm QM},i}}\right) \tag{4}$$

with

$$\frac{\partial E_{\rm QM}^{\rm cluster}}{\partial \vec{r}_{{\rm QM},i}} = \left(\frac{\partial E_{\rm QM}^{\rm cluster}}{\partial \vec{r}_{{\rm QM},i}}\right)_{{\rm fix}\,\{\vec{r}_{\rm b}\}} + \sum_j^{\rm buffer} \frac{\partial E_{\rm QM}^{\rm cluster}}{\partial \vec{r}_{{\rm b},j}}\left(\frac{\partial \vec{r}_{{\rm b},j}}{\partial \vec{r}_{{\rm QM},i}}\right)_{\rm CL} \tag{5}$$

and

$$\frac{\partial E_{\rm CL}^{\rm cluster}}{\partial \vec{r}_{{\rm QM},i}} = \left(\frac{\partial E_{\rm CL}^{\rm cluster}}{\partial \vec{r}_{{\rm QM},i}}\right)_{{\rm fix}\,\{\vec{r}_{\rm b}\}}. \tag{6}$$

The derivatives of the energies in Eqs. (5) and (6) are obtained analytically, while the last term $(\partial \vec{r}_{{\rm b},j} / \partial \vec{r}_{{\rm QM},i})_{\rm CL}$ for the pair (i,j) in Eq. (5) is evaluated through explicit numerical-differentiation for the buffered cluster in the CL calculation. For CL atom-i, the cluster terms, $E_{\rm CL}^{\rm cluster}$ and $E_{\rm QM}^{\rm cluster}$, do not contribute to the force; hence,

$$\vec{F}_i = -\frac{\partial E_{\rm CL}^{\rm system}}{\partial \vec{r}_{{\rm CL},i}}. \tag{7}$$

In the buffered-cluster method, positions of the buffer atoms in the atomic cluster for the QM calculation are determined using the relaxed positions of the buffer atoms obtained in the CL calculation. Therefore the computation times to determine positions of the buffer atoms for the QM calculation are negligible as compared to the DFT (*i.e.*, QM) calculation itself. If we were to minimize the DFT energy, $E_{\mathrm{QM}}^{\mathrm{cluster}}$, of the atomic cluster through relaxation of the buffer atoms in the QM calculation, there should appear various surface-reconstructions of the cluster, resulting in undesirable deformations of the crystalline structure. Such a reconstruction on the surface of the cluster is suppressed in the buffered-cluster method. We, in fact, apply the buffered-cluster method to crystalline Si with various choices of the QM region; the total number of the QM atoms ranges in 1 to 70. We use the hybridized Hamiltonian, Eq. (1), with the buffered-cluster method and obtain the ground-state configuration by moving all the atoms until the maximum atomic-force becomes smaller than 10^{-4}Ry/a.u. It is found that the maximum displacement of the atom from the crystalline position at the initial is about 0.03Å or smaller depending on the choice of the QM region.

The hybrid QM-CL simulation scheme with the buffered-cluster method will be applied to various exothermic processes in materials such as the crack propagation and the friction.[13] In the hybrid simulation run of such a process, the reaction region may move and increase in size as the simulation proceeds. Adaptive re-choosing of the QM region during the simulation run will be required for fast simulation. Since virtually no limitation exists in choosing the QM region in the buffered-cluster method, automatic re-choosing of the QM region is possible during the simulation run with simple algorithms. An example of the algorithm to be applied to the hybrid simulation about reaction of an H_2O molecule with a nanostructured Si has been explained in Sec. III.B of Ref. 8. The algorithm will be applied at a regular interval of 10fs in the simulation run.

The buffered-cluster method gives reasonable accuracies not only for covalent materials as Si in which the empirical inter-atomic potential is relatively short-ranged but also for ionic materials as alumina in which the empirical inter-atomic potential is long-ranged. In the hybrid QM-CL simulation[14] of the α-Al_2O_3 system with the buffered-cluster method, we adopt the variable-charge inter-atomic potential[15] for the CL calculation, in which the atoms have the effective charges determined dynamically following the generalized electro-negativity equalization principles. Proper coupling between the QM and the CL regions in the crystalline α-Al_2O_3 is realized with the buffered-cluster method when the QM region is chosen so that the outer QM-atoms (*i.e.*, the QM atoms connecting to the CL atoms) are Al. Double layers of the buffer atoms are set in the atomic clusters for the QM and the CL calculations. In the atomic cluster for the QM calculation, the first buffer-layer consists of O atoms; the second buffer-layer, of Al and "virtual" atoms. The virtual atom has the core charge $Z_{\mathrm{core}} = 0.5\,|e|$ with the non-integral number 0.5 of electrons. Positions of the buffer atoms are determined from that of the buffer atoms in the atomic cluster for the CL calculation. We note that reasonable accuracies of the buffered-cluster method in mechanical and electronical points of view have been confirmed for both Si and alumina systems.

In the link-atom method,[4-7] an H atom is put at the scaled position of the each atomic bond at the QM-CL boundary. The buffered-cluster method differs from the link-atom method in two crucial points: (i) Not only H but Si is used as the buffer atom in the case of the Si system. (ii) Positions of the buffer atoms are determined by using the relaxed configuration of the corresponding buffer atoms in the CL calculation, not by referring to the positions of the handshake atoms. Introduction of other species than H is necessary to reduce artificial deformation originating from mutual repulsion between the proximate H-atoms in the crystalline

Si system; for the α-Al_2O_3 system, H atoms cannot be used to terminate the broken bonds at the CL-QM boundary. Simple connection of the buffer-Si to the handshake-Si in the Si system would result in less accuracy in the mechanical coupling between the QM and the CL atoms as compared to the buffered-cluster method since the relaxed configuration of the buffer-Si differs substantially from the crystalline one and the strength of the bond between the buffer-Si and the QM-Si differs from that of the two QM-Si atoms.

In the buffered-cluster method, we have an option to saturate the buffer-Si further by H atoms for the Si system. For the α-Al_2O_3 system, however, the QM cluster cannot be saturated by, *e.g.*, H atoms. Therefore, the buffered-cluster method without such saturation may be considered as a simple and universal method applicable to both Si and α-Al_2O_3 systems.

We note that the long-range Coulomb interaction between the CL and the QM atoms is considered through the CL potential of the total system $E_{\mathrm{CL}}^{\mathrm{system}}$ (see, Eq. 1). For the Si system, the SW potential that we adopt has no atomic charges and hence the Coulomb interaction is ignored in practice. For the α-Al_2O_3 system, the Coulomb interaction is calculated with the atomic charges in the variable-charge potential that vary dynamically.

Practical implementation of the CG particle method

Recently, Rudd and Broughton[11] developed a new continuum approximation method for elastic materials. It is called the coarse-grained (CG) particle method. In the method, the CG particles are set into the target atomic-system and the interaction potential between the CG particles is obtained by taking constrained thermal average of the atomic system. Through phonon spectra analyses and wave reflection analyses it is shown that the CG particle method gives accurate results than the finite-element-method does. We note that the CG potential becomes equivalent to the atomic potential with the harmonic approximation when we set the CG particles on the atomic positions. Because of this feature, the CG region may connect to the atomic region seamlessly by gradually decreasing the degrees of coarsening near the region boundary. Therefore the CG particle method is expected to play a key role in developing a hybrid simulation code for large-scale realistic materials.

In the harmonic approximation, the total energy of the atomistic model is calculated as:

$$E^{\mathrm{total}} = \sum_{\alpha}^{N} \frac{\vec{p}_\alpha^{\,2}}{2m_\alpha} + \sum_{\alpha,\beta}^{N} \frac{1}{2} \vec{u}_\alpha \vec{D}_{\alpha,\beta} \vec{u}_\beta . \tag{8}$$

Here $\vec{u}_\alpha$ and $\vec{p}_\alpha$ are the displacement and the momentum of atom-α, respectively. The m_α is mass of atom-α. The $\vec{D}_{\alpha,\beta}$ represents the dynamical matrix of the atomistic system. Denoting displacement vector of the CG particle-*i* as $\vec{U}_i$, we set constraint relations for each particle-*i*:

$$\Delta = \prod_{i}^{N_{\mathrm{CG}}} \delta\left(\vec{U}_i - \sum_{\alpha}^{N} f_{i,\alpha}\vec{u}_\alpha\right) \delta\left(\dot{\vec{U}}_i - \sum_{\alpha}^{N} f_{i,\alpha}\frac{\vec{p}_\alpha}{m_\alpha}\right). \tag{9}$$

By performing statistical average of the total energy we obtain the total energy of the CG particles (Z is the partition function):

$$E^{\mathrm{CG}} = \frac{1}{Z}\left(\prod_{\alpha}^{N} \iint d\vec{u}_\alpha d\vec{p}_\alpha\right) E^{\mathrm{total}} \exp\left(-\frac{E^{\mathrm{total}}}{k_B T}\right) \Delta . \tag{10}$$

The CG particle method in its original formulation, summarized above, has drawbacks. Its direct application to a realistic nano-device is impossible since the matrix inverse of D, which appears after integration in Eq. (10), requires $O(N^3)$ operations. Also the inter-particle interaction formula cannot be applied directly to a system with translation, deformation, or rotation since each phonon-mode is constructed by combining the atomic displacements measured from the equilibrium positions with the reference axes fixed. We modify the CG particle method to overcome them as follows:

[Fast calculation of the stiffness matrix] We introduce the cutoff wavelength into the phonons and ignore those phonons with the wavelength longer than the cutoff value. This procedure corresponds to considering a repeating supercell of appropriate size for calculation of D. The size of the supercell should be chosen carefully considering both the computation speed and the physical accuracy required.

[The CG potential as a function of inter-particle separations] In the original formulation, the interaction potential between the CG particles is written as a function of their displacements not as a function of their absolute positions. This causes difficulty when the system deforms much. We have succeeded to reformulate the CG interaction potential as a function of CG positions.

[Adaptation to Rotation] In the original formulation, the CG interaction potential is written with rotation angle of the total system fixed. In real applications, the simulation system often rotates or deforms as the simulation proceeds. In the present CG method, local rotation-angles are determined dynamically, which are used to calculate the CG interaction potentials. This procedure makes us possible to apply the present CG method to realistic systems.

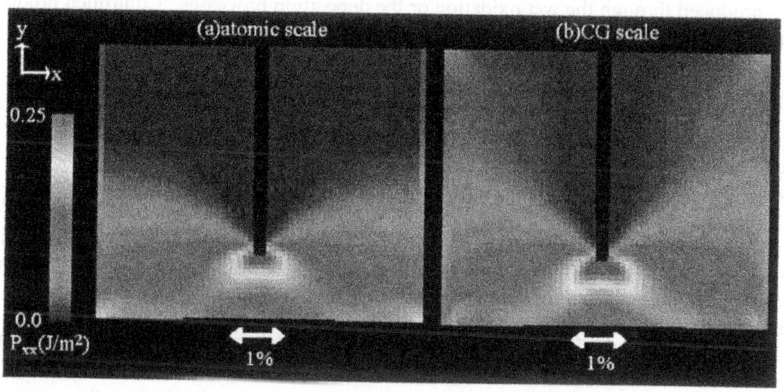

Fig. 2: Stress variations of the two dimensional Lennard-Jones system with a notch: (a) atomic analysis, (b) CG analysis.

To demonstrate high accuracy of the present CG particle method, we apply it to a two dimensional atomistic system in which atoms interact through the Lennard-Jones potential. System size is 1,400Å × 1,200Å with a notch at the center of the system. Number of atoms and CG particles are 128,164 and 3,555, respectively. The system is 1% stretched along the horizontal axis and is relaxed with the side-edges fixed. Figure 2 shows the result of stress distribution. Clearly the MD analysis and the CG analysis are quite in the same way.

APPLICATIONS

H_2O reaction on stressed Si-surfaces[8]

The hybrid QM-CL simulation scheme with the buffered-cluster method may be applied usefully to various phases of atomistic processes of engineering interests in nanostructured materials. The micro-electro-mechanical-system[16] (MEMS) is a complex nano-system composed of the electronic and the mechanical components in sub-micrometer ranges, which is made of a Si-based material and fabricated using the semiconductor processing techniques such as the directional etching. Despite growing expectation of its usage in sensing, control, and medical applications, the MEMS may exhibit unwanted failure due to the accelerated corrosion resulting from high-frequency stress-variations of the components. For instance, while the stress corrosion cracking is not observed for the bulk Si in an oxidizing environment, it is observed for Si-MEMS. Since the MEMS is expected to be used in mission-critical applications, its lifetime prediction is crucial. Theoretical understanding of chemical reactions of environmental molecules with nanostructured Si-systems under stresses will be useful in designing the MEMS.

Stress dependence of the oxidation reaction of Si is also a principal theme of active researches, with the intention of exploiting the information to realization of advanced computer and memory chips. A CMOS FET with the gate SiO_2-length on the order of 10^0nm, corresponding to several atomic layers, has been targeted. The gate dielectrics of the FET are usually produced through the wet oxidation or the deposition processes. Controlled production of such an ultra-thin oxide layer requires detailed knowledge about the dissociation and oxidation processes of water molecules on the nanostructured Si under various external stresses relating to the three-dimensional fine structures of the system.

Motivated by these, we study the initial stages of H_2O-reaction with a nanostructured Si using the hybrid QM-CL simulation scheme with the buffered-cluster method. The initial stages include adsorption and dissociation processes of an H_2O molecule on the Si system, which may vary qualitatively depending on the degree of strain applied to the system.

We prepare a Si-slab system with dimensions (L_x, L_y, L_z) =(108.6Å, 27.1Å, 70.6Å). The periodic boundary conditions are assumed in the x and y directions, while the free boundary condition in the z direction. The diamond crystalline-structure is assumed to the system; the x-axis corresponds to $<1\bar{1}0>$ direction, the y-axis to $<110>$, and the z-axis to $<001>$. Then a notch with x-width ~16Å and z-height ~36Å is inserted at the middle of the system. The bottom of the notch assumes a facet with (001) structure, and two sidewalls of the notch with $(1\bar{1}0)$ structure. Additionally, facets with (111) structure are created at the corners between the bottom and the two sidewalls of the notch, which act to stabilize the initial crystalline configuration around the corners. We prepare another system by removing one more atomic-layer from the bottom of the notch in the Si-slab system explained above. In both systems, we initially create symmetric (1×1) Si-Si dimers on the bottom of the notch, and then perform MD simulation with the velocity dumping method using the SW inter-atomic potential to obtain the relaxed atomic-configuration. The dimers are perpendicular to the x direction in one system, while they are perpendicular to the y direction in the other system.

To prepare stretched systems we firstly expand both systems uniformly by 4% in the x direction and then relax their atomic positions by the MD simulation with the velocity dumping method until the maximum atomic-force becomes smaller than 10^{-4}Ry/a.u. To characterize stress distributions in the stretched systems, we draw the atomic virial-stresses σ_{xx} in the systems in Fig. 3: (a) for the system with the dimers oriented perpendicularly to the x direction; (b) for the system, to the y direction. In the stretched systems, stress concentration is observed at the bottom of the notch. The maximum magnitude of σ_{xx} is about 8GPa at the center of the notches in the stretched system, which is about four times larger than the averaged magnitude of 2GPa at the low-x and the high-x ends of the system. Interestingly stress variations near the bottom of the notch differ significantly between the two systems as seen in Figs. 3(a) and 3(b), which is a reflection of different combinations of the stretch and the dimer directions between the two systems.

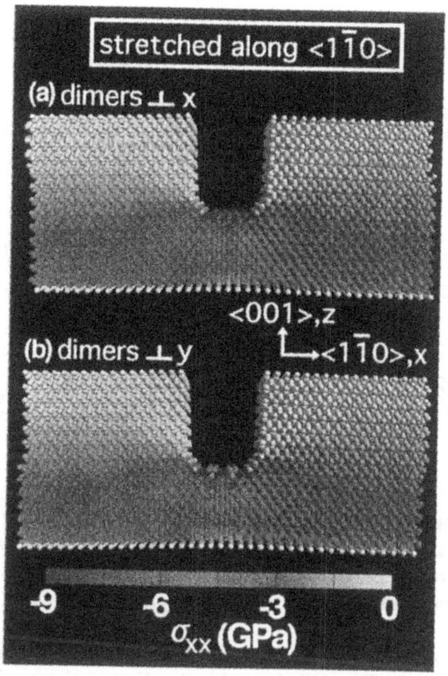

Fig. 3: The atomic virial-stresses σ_{xx} in the 4%-stretched Si system with the notch obtained with the classical MD simulation. (a) For the system with Si-Si dimers at the bottom of the notch oriented perpendicularly to the x direction. (b) For the system with Si-Si dimers at the bottom of the notch oriented perpendicularly to the y direction.

We depict in Fig. 4 three dissociation processes of the adsorbed H_2O at the bottom-facet of the notch with Si(100)-(2×1) dimer structure in the unstrained Si-system. Configuration of the QM atoms chosen in the hybrid simulation is depicted in Fig. 4(a) for the adsorbed state. The three processes considered are: dissociation on the same dimer, that is, from (a) to (b) in Fig. 4; dissociation to the first nearest-neighbor dimer, from (a) to (c); and dissociation to the second nearest-neighbor dimer, from (a) to (d). Barrier energies and dissociation energies Q for the dissociation process are written in Fig. 4.

From the thermodynamic point of view, small barrier and large dissociation-energy, Q, are favorable for the dissociation process to occur. However, we remind that the adsorbed-H_2O may have the kinetic energy of 0.7eV corresponding to the adsorption energy,[17] which is larger than or nearly equal to the barriers of 0.15, 0.4, and 0.6eV for the three dissociation processes considered. We may expect that all the three dissociation processes may occur depending on the initial settings.

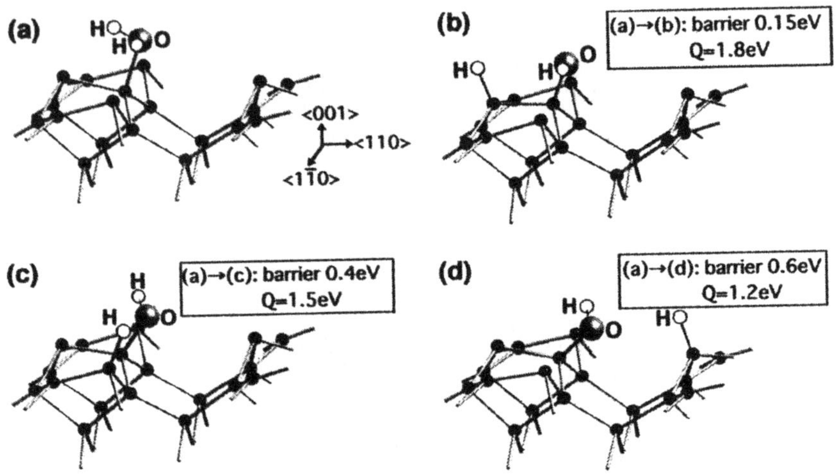

Fig. 4: (a) Atomic configuration of H_2O-adsorbed Si(100) facet in the unstrained Si-system. (b) The dissociated state with H and OH adsorbed to the same Si-Si dimer. (c) The dissociated state with H adsorbed to the first nearest-neighbor dimer. (d) The dissociated state with H adsorbed to the second nearest-neighbor dimer.

We perform the hybrid QM-CL simulation runs of the H_2O reaction with the notched Si-system with or without strains. We will show that the dissociation to the second-nearest neighbor dimer in fact occurs in the unstrained system. The QM region chosen in the system contains 33 Si atoms and a single H_2O molecule. The H_2O is set above the down atom of a Si-Si dimer on the bottom-facet of the notch with $r_{Si\text{-}O} = 3.1Å$. Positions of all the atoms are relaxed with the constraint of $r_{Si\text{-}O} = 3.1Å$.

For the unstrained system with the notch, we perform five simulation runs with different initial velocities for the H_2O molecule. We set the same velocity $\vec{v}$ to 2H and O atoms of the

H_2O; magnitude of $\bar{v}$ corresponds to temperature $T = 600K$ of the O atom, *i.e.*, $\frac{1}{2}m_O v^2 = \frac{3}{2}k_B T$. The initial velocities in the five runs are $(v_x, v_y, v_z) =$ (0m/s,0m/s,-680m/s) , (0m/s,-200m/s,-650m/s) , (0m/s,200m/s,-650m/s) , (−200m/s,0m/s,-650m/s) , and (200m/s,0m/s,-650m/s). Since the O and H atoms move during the simulation run, expansion of the QM region becomes necessary. We apply the algorithm explained in Methodology section at every 10fs to re-choose the QM atoms adaptively in the total system. The time step is 0.67fs. No temperature control is performed. In two of the five runs, we find that the H_2O dissociates to H and OH and that H adsorbs to a dangling-bond site of the second nearest-neighbor dimer, which corresponds to the process from (a) to (d) in Fig. 4. In other three runs, the H_2O adsorbs simply to a dangling-bond site of the dimer without dissociation.

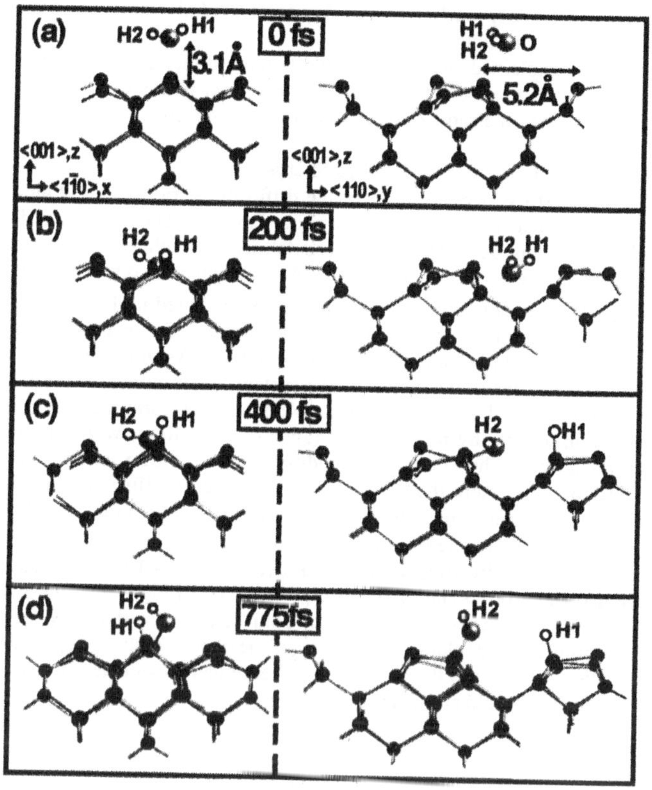

Fig. 5: Time evolution of configuration of the QM atoms in the hybrid QM-CL simulation of the dissociation process of an H_2O in the unstrained Si-system with the buffered-cluster method.

Time evolution of the atomic configuration in the dissociation run of the unstrained system with $(v_x,v_y,v_z)=$(0m/s,0m/s,-680m/s) is depicted in Fig. 5. The QM region expands at 40, 70, 210, 390, and 450fs with addition of 2, 1, 3, 3, and 3 QM-Si atoms, respectively, as the simulation proceeds.

We investigate dependence of the reaction dynamics of the H_2O molecule with the notched Si-system on the strain applied to the system. We consider two 4%-stretched Si-systems with the notch; dimers on the bottom of the notch are perpendicular to x direction in one system, while they are perpendicular to y direction in the other system. Initially the H_2O molecule is set at $r_{\text{Si-O}}=3.1$Å above the down atom of a dimer, in a similar way to the case of the unstrained Si-system. For each system, we perform five simulation runs with the same initial velocities of the H_2O molecule as that in the simulation runs for the unstrained system. In all the runs, we find that the H_2O molecule approaches to the down atom of the dimer and dissociates, and that resulting H and OH adsorb to the same dimer.

As we have explained, we find that the H atom dissociated from the H_2O molecule has substantial probabilities to adsorb to the neighboring dimer in the unstrained Si systems, and high probabilities to adsorb to the same dimer in the stretched systems.

Diffusion barriers for an O atom in stretched Si crystals

We consider stress dependencies of the diffusion barrier for an O atom in crystalline Si. In unstrained Si, the barrier is known to be about 2.3eV.[18] Data on stress dependencies of the barriers will be useful in developing advanced fabrication techniques for nano-scale interfaces. The total system that we consider contains 36,864 Si atoms with dimensions $(L_x,L_y,L_z)=$(92.2Å, 92.2Å, 86.9Å). The periodic boundary conditions are assumed in the x, y, and z directions. The diamond crystalline-structure is assumed to the system; the x-axis corresponds to $<1\bar{1}0>$ direction, the y-axis to $<110>$, and the z-axis to $<001>$. As shown in Fig. 6 (top), we select 13 Si-atoms as the QM atoms at the center of the system; other atoms are the CL atoms. Then we put an O atom to be treated as the QM atom. We thereby perform the hybrid QM-CL simulation to obtain the oxidized state by relaxing the atomic configurations with subtraction of the kinetic energy of all the atoms. The O atom is found to sit between Si-5 and Si-2 atoms as depicted in the inset of Fig. 6. Separately the transition state for the O diffusion is obtained via the hybrid QM-CL simulation by controlling angles $\theta_{\text{O}}=90^\circ$ and $\theta_{\text{Si}}=270^\circ$ in Fig. 6 following Ref. 18

Figure 7 shows calculated barrier values as functions of the percent of stretch applied to the total system. Three directions of stretch are considered: $<1\bar{1}0>$ (i.e., x direction), $<110>$ (i.e., z direction), and $<001>$ (i.e., y direction). We find in Fig. 7 that the stress dependence of the barrier varies significantly depending on both direction and degree of stretch. Detailed analyses are in progress to explain mechanisms of the observed stress dependencies. The hybrid QM-CL simulation scheme has the capability to theoretically evaluate diffusion barrier for a trace atom in a realistic stress-field at the nano-interface.

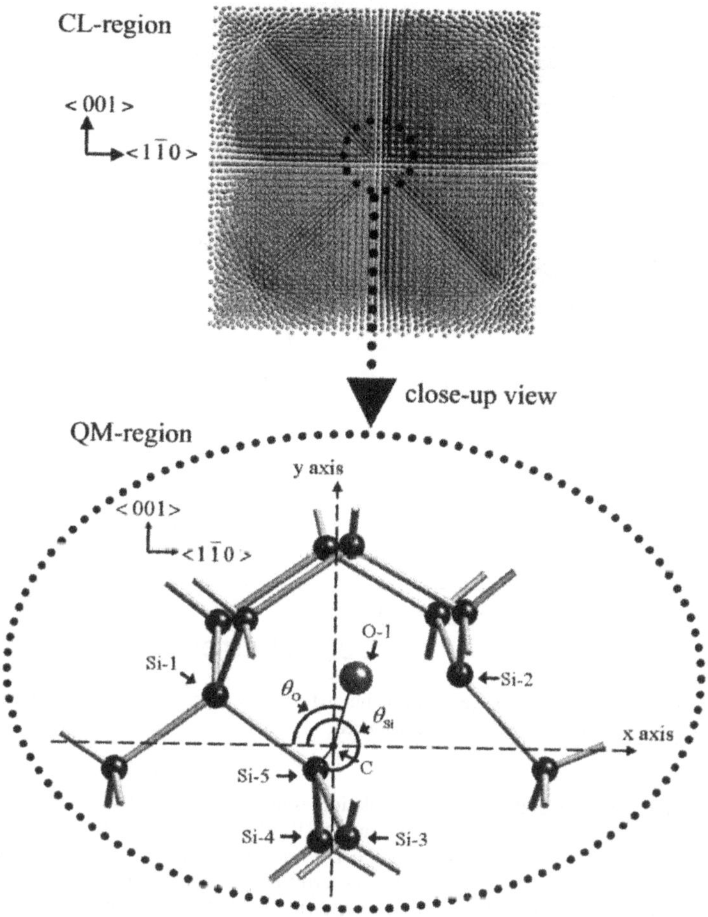

Fig. 6: (Top) Total system of crystalline Si for the hybrid QM-CL simulation, to calculate diffusion barrier for an O atom. (Inset) Close-up view of the QM region.

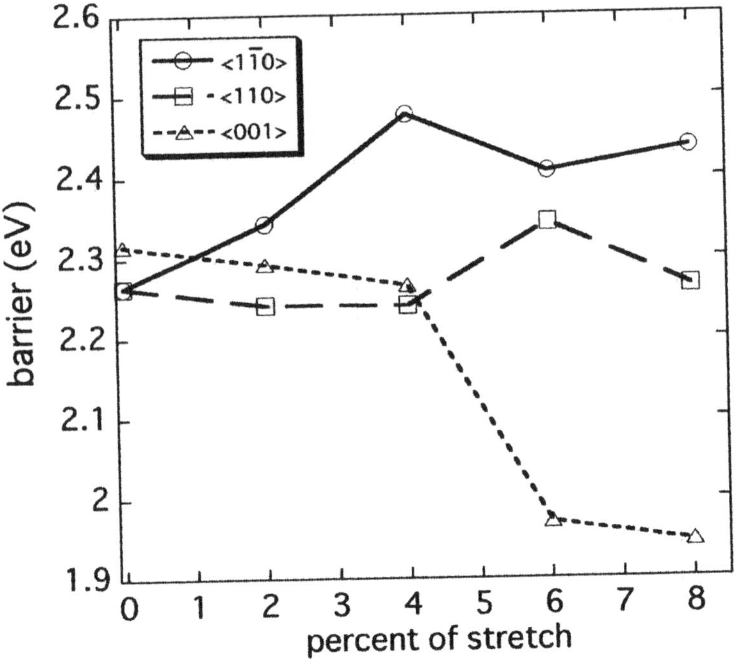

Fig. 7: Hybrid QM-CL simulation results for the diffusion barrier for an O atom in crystalline Si. Three different directions of stretch are considered.

SUMMARY

We have developed the hybrid simulation schemes for dynamic simulations of such processes on computation grids, by combining DFT, MD, and CG particle methods. In the hybrid scheme, a QM region treated with the DFT is embedded in a large-scale CL system composed of the MD atoms. The CG particle description gives the dynamic boundary condition to the CL region. We proposed the buffered-cluster method for robust connection of the QM and CL regions, with introduction of the buffer atoms to minimize possible effects arisen from finiteness of the size of the QM region. The buffered-cluster method is applicable to any reasonable choice of the QM region in a wide range of ceramics and semiconductor materials. Insensitivity of the atomic forces to the choice of the QM region in the buffered-cluster method makes it possible to re-choose the QM region adaptively during the hybrid simulation run for fast computation. The hybrid QM-CL simulation scheme with the buffered-cluster method was used successfully to analyze adsorption and dissociation processes of an H_2O molecule on a notched Si-slab system with or without strains and to evaluate stress-dependent diffusion barriers for an O atom in Si crystals.

ACKNOWLEDGEMENT

This work was supported in part by JST-CREST and the Grant-in-Aid for Scientific Research of the Ministry of Education, Science, Sports, and Culture of Japan

REFERENCES

[1]A. Nakano, M.E. Bachlechner, P. Branicio, T.J. Campbell, I. Ebbsjö, R.K. Kalia, A. Madhukar, S. Ogata, A. Omeltchenko, J.P. Rino, F. Shimojo, P. Walsh, and P. Vashishta, "Large Scale Atomistic Modeling of Nanoelectronic Structures," IEEE Trans. Elec. Dev. 47, 1804-1811 (2000).

[2]A. Nakano, M.E. Bachlechner, R.K. Kalia, E. Lidorikis, P. Vashishta, G.Z. Voyiadjis, T.J. Campbell, S. Ogata, and F. Shimojo, "Multiscale Simulation of Nanosystems," Comp. in Sci. & Eng. 3, 56-65 (2001).

[3]G. Lu and E. Kaxiras, in *Handbook of Theoretical and Computational Nanotechnology, Vol. X*, edited by M. Rieth and W. Schommers (Amer. Sci. Pub., NY, 2005), Chap. 22.

[4]S. Dapprich, I. Komáromi, K.S. Byun, K. Morokuma, and M.J. Frisch, "A new ONIOM Implementation in Gaussian98. Part 1. The calculation of energies, gradients, vivrational frequencies and electric field derivatives," *J. Mol. Struc. (Theochem)* 461-462, 1-21 (1999).

[5]M. Eichinger, P. Tavan, J. Hutter, and M. Parinello, "A Hybrid Method for Solutes in Complex Solvents: Density Functional Theory Combined with Empirical Force Fields," *J. Chem. Phys.* 110, 10452-10467 (1999).

[6]S. Ogata, E. Lidorikis, F. Shimojo, A.Nakano, P. Vashishta, and R.K. Kalia, "Hybrid Finite-Element/Molecular-Dynamics/Electronic-Density-Functional Approach to Materials Simulations on Parallel Computers"*Comp. Phys. Comm.* 138, 143-154 (2001).

[7]S. Ogata, F. Shimojo, R.K. Kalia, A. Nakano, and P. Vashishta, "Hybrid Quantum Mechanical/Molecular Dynamics Simulations for Parallel Computers: Density Functional Theory on Real-Space Multigrids,"*Comp. Phys. Comm.* 149, 30-38 (2002).

[8]S. Ogata, "Buffered-Cluster Method for Hybridization of Density-Functional Theory and Classical Molecular Dynamics: Application to Stress-Dependent Reaction of H_2O on Nanostructured Si", *Phys. Rev. B* 72, 45348-45364 (2005)

[9]e.g, M.C. Payne, M.P. Teter, D.C. Allan, T.A. Arias, and J.D. Joannopoulos, "Iterative Minimization Techniques for ab Initio Total-Energy Calculations: Molecular Dynamics and Conjugate Gradients", *Rev. Mod. Phys.* 64, 1045-1097 (1992).

[10]S. Ogata, F. Shimojo, A. Nakano, P. Vashishta, and R.K. Kalia, "Environment effects of H_2O on fracture initiation in Si: a hybrid electronic-density-functional/molecular-dynamics study," *J. Appl. Phys.* 95, 5316-5323 (2004).

[11]R.E. Rudd and J.Q. Broughton, "Coarse-Grained Molecular Dynamics and the Atomic Limit of Finite Elements," *Phys. Rev. B* 58, R5893-R5896 (1998).

[12]F.H. Stillinger and T.A. Weber, "Computer Simulation of Local Order in Condensed Phases of Silicon," *Phys. Rev. B* 31, 5262-5271 (1985).

[13]E. Meyer, R.M. Overney, K. Dransfeld, and T. Gyalog, *Nanoscience: Friction and Rheology on the Nanometer Scale* (World Scientific, Singapore, 1999), p. 349.

[14]S. Ogata and R. Belkada, "A Hybrid electronic-density-functional/molecular-dynamics simulation scheme for multiscale simulation of materials on parallel computers: application to silicon and alumina," *Comp. Mat. Sci.* 30, 189-194 (2004).

[15]F.H. Streitz and J.W. Mintmire, "Electrostatic Potentials for Metal-Oxide Surfaces and Interfaces," *Phys. Rev. B* 50, 11996-12003 (1994).

[16]IEEE, *Microelectromechanical Systems (MEMS), 1999 IEEE 12th International* (IEEE, NY, 1999).
[17]K. Akagi and M. Tsukada, "Theoretical Study of the Hydrogen Relay Dissociation of Water Molecules on Si(001) Surfaces," *Surf. Sci.* 438, 9-17 (1999).
[18]M. Ramamoorthy and S.T. Pantelides, "Coupled-Barrier Diffusion: The Case of Oxygen in Silicon," *Phys. Rev. Lett.* 76, 267-271 (1996).

MOLECULAR DYNAMICS SIMULATIONS OF THE EFFECT OF THE COMPOSITION OF INTERGRANULAR FILMS ON GRAIN GROWTH

Stephen H. Garofalini
Interfacial Molecular Science Laboratory
Dept of Materials Science and Engineering
Rutgers University
Piscataway, NJ, 08855

ABSTRACT

Molecular dynamics simulations were performed to study the effect of the composition of the intergranular film (IGF) on grain growth in α-Al_2O_3 and β-Si_3N_4. In each case, the IGF is formed while in contact with two differently oriented crystals. Results show preferential growth along the $[11\bar{2}0]$ direction of the $(11\bar{2}0)$ surface in comparison to growth along the surface normal on the (0001) surface for certain calcium alumino-silicate compositions in the alumina system. However, in the silicon nitride system, preferential growth occurred along the surface normal on the (0001) surface in comparison to that in the $(10\bar{1}0)$ surface. Such preferential growth in each system is consistent with experimentally observed anisotropic grain growth in each of the different systems. Platelets form in the alumina system because of faster outward growth of the prism orientations than the basal orientation, while the more rapid growth along the basal surface normal in the nitride system causes anisotropic growth to manifest itself as whisker growth. In the alumina system, the simulations show the mechanism by which Ca ions in the IGF inhibit growth on the basal surface and the Ca/Al ratio in the IGF plays an important role in the change from isotropic to anisotropic grain growth. At compositions with high or low Ca/Al ratios, growth along each surface normal is equivalent, indicating isotropic grain growth. The simulations provide an atomistic view of attachment onto crystal surfaces, affecting grain growth in alumina.

INTRODUCTION

Amorphous intergranular films often form between crystals in commercial ceramics because of the presence of impurities present in the materials or because of the addition of sintering aids[1-8]. These films affect microstructural and mechanical properties. Ultrapure Al_2O_3 sintered in clean conditions usually shows normal grain growth[9, 10], while commercial alumina usually contains sufficient impurities that modify microstructure, via anisotropic grain growth, and resultant properties. This anisotropic growth forms platelet crystals that have large, flat basal surfaces with thin intergranular films[11]. Silicon nitride requires sintering aids to enable densification during sintering. These aids create silicate IGFs that also affect the microstructure and properties of the nitride[4, 12-14]. In the nitride case, anisotropic grain growth causes formation of whiskers rather than platelets. In both cases, the IGF appears to affect grain growth, with the implication that the atoms in the IGF affect adsorption and growth along the surface normal of specific planes.

Experiments showed that small (ppm) levels of either calcia or silica in the cause abnormal grain growth (AGG) in alumina[15]. Understanding the atomistic structure of the IGF in the polycrystalline ceramics is experimentally difficult because of the thin and amorphous structures of these films, although there has been recent success in studies of IGFs in Si_3N_4 [16-18]. In concert with experiment, computational studies have been used to address these films at the atomistic level[19-33].

Our previous simulations of amorphous silicate IGFs in contact with alumina crystals and silicon nitride crystals clearly showed ordered structures at the interfaces and density oscillations into the IGF induced by the crystals[20-22, 29, 30, 34]. Recent High Resolution TEM (HRTEM) studies observe such density oscillations[18]. These density oscillations decay within a nanometer of the interface, attaining a glassy structure in the interior of the thicker films. Most importantly, the previous simulations of IGFs between the differently oriented crystals showed anisotropic growth[30, 34]. The term 'growth' in this discussion means adsorption of species from the IGF onto the crystal surface consistent with the crystal orientation and composition such that the crystal lattice extends along the surface normal of the particular plane in contact with the IGF. In a real system, such adsorption of species from the IGF would require the neighboring crystal to move away as the adsorbing crystal extended outward if the separation distance between the crystals must remain nearly constant. Also, the adsorbing species from the IGF must be replenished if growth is to continue. In the simulations, the use of constant pressure along the surface normal enables the crystals to move according to the structural changes. However, we do not replenish the IGF in the current simulations, although the use of a variety of compositions provides useful information regarding the effect of changing composition on adsorption and growth behavior. In the nitride case, simulations showed that growth along the basal surface normal was more rapid than that along the prism surface normal. In the alumina case, the simulations showed preferential growth along the $[11\bar{2}0]$ direction on the $(11\bar{2}0)$ oriented crystal and no such growth in the $[0001]$ direction on the basal crystal. Both results are consistent with anisotropic grain growth in the respective systems seen experimentally.

COMPUTER PROCEDURE

Similar to previous studies[19-21, 34-36], a multibody interatomic potential was used in the simulations and is given as:

$$V = \sum_{i \neq j} V_{ij}^{BMH} + \sum_{i \neq j \neq k} V_{jik}^{3-body} \qquad (1)$$

where V_{ij}^{BMH} is the pairwise modified Born Mayer Huggins (BMH) potential term and V_{jik}^{3-body} is the three body potential term. The modified BMH potential is defined as

$$V_{ij}^{BMH} = A_{ij} \exp\left(\frac{-r_{ij}}{\rho_{ij}}\right) + \frac{z_i z_j e^2}{r_{ij}} \xi \left(\frac{r_{ij}}{\beta_{ij}}\right) \qquad (2)$$

The first term is the repulsion term in the modified BMH pair potential and represents the core electron overlap when the two atoms get too close; the second term is a screened Coulomb interaction which reduces the nominal charges based on ξ (which is a complimentary error function). r_{ij} is the separation distance between the ions i and j, z_i and z_j are the formal charge of the ions and β_{ij} is a species dependent term that reduces the formal charges as a function of distance between the ij pair. The value of the parameters A_{ij}, β_{ij} and ρ_{ij} for each pair type were previously presented[20, 30, 32, 34, 35]. Atomic charges for silicon, oxygen, nitrogen, aluminum and calcium ions are +4, -2, -3, +3 and +2, respectively.

The partial covalent bonding of the species has preferred bonding angles at which the bonded species have the lowest energy. When the bonding angle deviates from the preferred one, the energy will increase. The 3-body potential accounts for this effect by raising the energy on a central ion within a triplet as the angle deviates from an ideal angle. The 3-body potentials applied to all Si, O, N, and Al ions as central ions i is given as:

$$V_{jik}^{3-body} = \lambda_{jik} \exp \frac{\gamma_{ij}}{r_{ij} - r_{ij}^{0}} + \frac{\gamma_{ik}}{r_{ik} - r_{ik}^{0}} \Omega_{jik} \qquad (3)$$

when $r_{ij} < r_{ij}^{0}$ and $r_{ik} < r_{ik}^{0}$.

Otherwise,

$V_{jik}^{3-body} = 0$ when $r_{ij} > r_{ij}^{0}$ or $r_{ik} > r_{ik}^{0}$.

The angular part, Ω_{jik}, is given as

$$\Omega_{jik} = \left(\cos\theta_{jik} - \cos\theta_{jik}^{0}\right)^{2} \qquad (4)$$

for the Si/Al/Ca-**O/N**-Si/Al/Ca and O/N-**Si**-O/N structure, while for the O-**Al**-O triplet, it is defined as

$$\Omega_{jik} = \left[\left(\cos\theta_{jik} - \cos\theta_{jik}^{0}\right)\sin\theta_{jik}\cos\theta_{jik}\right] \qquad (5)$$

where the three atoms form angle θ_{jik} with the central atoms as the vertex i in these triplets for both conditions. Since Ca is considered mainly ionic, no 3-body potential centered on Ca as ions *i* in the *jik* triplet is used. The parameters were also developed and used in previous studies[21, 30, 32, 35, 37, 38] and are given in Tables I and II .

In the alumina case, the glassy IGFs were made between two dissimilar α-alumina orientations, the (0001) and $(11\bar{2}0)$, creating a ((0001) oriented crystal)/(IGF)/($(11\bar{2}0)$ oriented crystal) system. In the nitride case, the orientations involved the (0001) and $(10\bar{1}0)$ surfaces with a calcium silicon oxynitride IGF between the crystals. Compositions and system sizes are given in Tables III and IV. A schematic drawing of the system is shown in figure 1. The simulation procedure is the same as that previously published[31, 34], where it has been discussed in detail.

Briefly, each crystal is aligned such that the appropriate crystal terminating plane would be in contact with the IGF, as shown in figure 1. The atoms in top few layers of the upper crystal, and the bottom layers of the lower crystal are always frozen so that periodic boundaries can be used in all three dimensions in the alumina case, since the frozen atoms never see each other (but can act as neighbors to the mobile atoms). The vacuum in the nitride case also acts to prevent the crystals from interacting across the periodic boundary in z. These frozen sections move en-masse to any imposed pressure.

The glassy IGFs are made between the crystals by randomly inserting the atoms of the IGF composition in the volume between the crystals, followed by a melt/quench procedure. The melt/quench process for making the glassy IGF between the crystals is shown in Table V for each system. Both NVE and NPT simulations are used and a timestep of $1x10^{-15}$s is used throughout the simulations. The x, y, and z dimensions are approximately 5.2nm, 5nm, 10nm, respectively. The x and y dimensions change with the thermal expansion coefficient of the crystal to minimize elastic effects. The IGF is ~4nm in the alumina simulations and 1-2nm in the nitride simulations. Again, details have been presented elsewhere [31, 34].

RESULTS AND DISCUSSION

Figure 2 shows the density profiles as a function of z for the atoms in the basal and prism oriented crystals and the sum of the atoms in the IGF the 112 alumina system. There are density oscillations in the IGF near the interfaces, decaying within ~1nm from each interface, showing a glassy density profile in the interior of the IGF. Similar density oscillations are observed in the nitride system, although the thinner dimension of the IGF in the nitride means that these density oscillations impinge on each other. Such density oscillations have been observed in recent aberration-corrected high resolution TEM[18, 39]. Figure 3 shows the density profiles as a function of z for the individual species for the alumina system with the anorthite calcium alumino-silicate IGF composition, in which the Al and O are labeled separately in the IGF versus in the crystals. Extension of the crystals along the surface normal in each case, implying growth along the surface normal, would require adsorption of Al ions from the IGF onto the specific crystal surface. The arrows indicate the first layer of Al adsorbed from the IGF onto the crystal (or onto adsorbed O from the IGF that have adsorbed onto the crystal in the prism side). The adsorbed Al peak is necessary for growth and can be used to indicate the potential for growth of the specific orientations as a function of composition in the IGF. The area under the first adsorbed Al peak, normalized by the area under the peak of the Al in the crystal for each appropriate crystal orientation, will be used as a measure of the 'growth' of each crystal along its surface normal.

Figure 4 shows the normalized first adsorbed Al peak for the basal and prism orientations for the alumina case as a function of the Al concentration in the IGF (normalized by the sum of the cations in the IGF). Clearly, adsorption and growth of the alumina is more rapid on the prism surface than on the basal surface, indicating anisotropic grain growth. This anisotropy is consistent with experimentally observed anisotropic grain growth in alumina. At higher Al concentrations, adsorption on the basal surface increases to the point where growth along each surface normal is similar, indicating isotropic grain growth.

The different adsorption and growth behavior on the different surfaces is caused by the presence of Ca in the IGF. Figure 5 shows a snapshot of a side view of the basal oriented crystal, with a 'cage' structure at the interface with the IGF. Ca ions preferentially adsorb into the cage interiors, inhibiting adsorption onto the crystal sites to continue the crystal structure in this orientation. One Ca ion charge compensates 2 Al in the tetrahedral sites in the glassy silicate structure. With excess Ca in the IGF, Ca ions segregate to the interface to charge compensate undercoordinated O at the crystal surface (and the zeolitic oxygen in the Al-O-Si sites at the 'cage' sides). With sufficiently high concentrations of Al in the IGF (in excess of the compensating Ca concentration), Al ions epitaxially adsorb onto the crystal on the basal side, extending the crystal in this direction.

On the prism side, increasing amounts of Al ions adsorb onto the surface with increasing concentration of Al in the IGF. There is a competition for these adsorption sites with the Si and Ca from the IGF.

Additional data shows that the role of the Ca/Al ratio plays an important role in the change in adsorption onto the basal surface[40]. At high and low Ca/Al ratios is the IGF, adsorption and growth are similar on both types of orientations, which may be indicative of isotropic growth, with a clear separation of peak intensities and adsorption behavior near the Ca/Al ratio equal to 0.5, which is indicative of anisotropic growth.

In the nitride case, N adsorbed from the calcium silicate oxynitride IGFs is the indicator of growth of the crystal along the surface normal for the basal and prism oriented crystals. Similar to the alumina case, the first peak of N epitaxially adsorbed from the IGF onto the crystal surfaces is normalized by the N peak in each specific crystal orientation. Results are shown in figure 6. In the nitride case, growth along the basal surface normal is more rapid than that on the prism surface normal, indication anisotropic growth. Experimentally, anisotropic growth in silicon nitride shows more rapid growth along the basal surface normal than the prism surface normal, consistent with these simulation results. This anisotropic growth forms whiskers in the nitride morphology.

Therefore, while the oxide and nitride systems show different types of anisotropic growth, the simulations accurately reproduce these different trends in growth behavior. The simulations were not biased to create these results. The potentials are consistent in each system and between systems. The interatomic potential for Si-O in the alumina case is the same as that in the nitride case; however, environment (other neighbors) does effect bond lengths and energies, as expected.

CONCLUSIONS

Experimentally, the oxide and nitride systems studied here show different anisotropic morphology (platelet in the former, whiskers in the latter system). The molecular dynamic simulations used here reproduce this different anisotropy without biasing the potentials to do so. The simulations show the effect of the composition of the intergranular film (IGF) between dissimilarly oriented α-Al_2O_3 crystals and β-Si_3N_4 crystals on growth in the direction of each surface normal. Results show anisotropic adsorption and growth at certain IGF compositions and more isotropic growth at other compositions.

During the kinetic evolution of a system during liquid phase sintering, the composition of the IGF is expected to change in time before reaching an equilibrium composition and thickness. The simulations show that these different compositions affect adsorption of ions from the IGF onto the different crystallographic orientations, subsequently affecting growth and morphology.

Acknowledgement: The author acknowledge support from the DOE OBES, Division of Materials Sciences, grant number DE-FG02-00ER45823 and partial support from NSF Award DMR-0010062 in collaboration with the NANOAM project of EU-CORDIS (G5RD-CT-2001-00586).

REFERENCES

1. W. A. Kaysser and M. Sprissler. "Effect of a Lquid Phase on the Morphology of Grain Growth in Alumina". *J. Am. Ceram. Soc.* **70**, [5] 339-343 (1987).
2. C. A. Powell-Dogan and A. H. Heuer. "Microstructure of 96% Alumina Ceramics: I, Characterization of the As-Sintered Materials". *J. Am. Ceram. Soc.* **73**, [12] 3670-76 (1990).
3. C. A. Powell-Dogan, A. H. Heur and H. M. O'Bryan. "Devitrification of the Grain Boundary Glassy Phase in a High-Alumina Ceramic Substrate". *J. Am. Ceram. Soc.* **77** 2593 (1994).
4. H.-J. Kleebe. "Structure and Chemistry of Interfaces in Si3N4 Ceramics Studied by Transmission Electron Microscopy". *J. Ceram. Soc. Japan* **105** 453-475 (1997).
5. H.-J. Kleebe, M. K. Cinibulk, R. M. Cannon and M. Rühle. "Statistical Analysis of the Intergranular Film Thickness in Silicon Nitride Ceramics". *J. Am. Ceram. Soc.* **76**, [8] 1969-1077 (1993).
6. G. Pezzotti, et al. "Chemistry and inherent viscosity of glasses segregated at grain boundaries of silicon nitride and silicon carbide ceramics". *Journal of Non-Crystalline Solids* **271** 79-87 (2000).
7. M. K. Cinibulk, H.-K. Kleebe and M. Rühle. "Amorphous intergranular films in silicon nitride ceramics quenched from high temperatures". *J. Am. Ceram. Soc.* **76**, [11] 2801-2808 (1993).

8. P. F. Becher, et al. "Debonding of Interfaces Between Beta-Silicon Nitride Whiskers and Si-Al-Y Oxynitride Glasses". *Acta. Mater.* **44** 3881-3893 (1996).
9. S. I. Bae. "Sintering and Grain Growth of Ultrapure Alumina". *J. of Mat. Sci.* **28** 4197-4204 (1993).
10. J.-S. Bae and S.-I. Pyun. "Electrochemical lithium intercalation reaction of anodic vanadium oxide film". *J. Alloys and Compnds* **217** 52-58 (1995).
11. C. W. Park. "Singular Grain Boundaries in Alumina and Their Roughening Transition". *J. Am. Ceram. Soc.* **86**, [4] 603-611 (2003).
12. F. F. Lange. "Fabrication and Properties of Dense Polyphase Silicon Nitride". *Am. Ceram. Soc. Bull.* **62**, [12] 1368-74 (1983).
13. R. L. Tsai and R. Raj. "The Role of Grain Boundary Sliding in Fracture of Hot-Pressed Si3N4 at High Temperatures". *J. Am. Ceram. Soc.* **63**, [9-10] 513-17 (1980).
14. R. Raj. "Creep in Polycrystalline Aggregates by Matter Transport Through a Liquid Phase". *J. Geophys. Res. B* **87**, [7] 4731-39 (1982).
15. S. I. Bae and S. Baik. "Determination of Critical Concentrations of Silica and/or Calcia for Abnormal Grain Growth in Alumina". *J. Am. Ceram. Soc.* **76**, [4] 1065-1067 (1993).
16. A. Ziegler, C. Kisielowski, M. J. Hoffmann and R. O. Ritchie. "Atomic resolution transmission electron microscopy of the intergranular structure of a Y2O3-containing silicon nitride ceramic". *J. Am. Ceram. Soc.* **86**, [10] 1777-1785 (2003).
17. N. Shibata, et al. "Observation of rare-earth segregation in silicon nitride ceramics at subnanometer dimensions". *Nature* **428** 730-733 (2004).
18. M. Doblinger, et al. "Structural and compositional comparison of Si3N4 ceramics with different fracture modes". *Acta Mater.* (in press).
19. S. Blonski and S. H. Garofalini. "Molecular Dynamics Study of Silica-Alumina Interfaces". *J. Phys. Chem.* **100** 2201-2205 (1996).
20. S. Blonski and S. H. Garofalini. "Atomistic Structure of Calcium Silicate Intergranular Films in Alumina Studied by Molecular Dynamics Simulations". *J. Am. Ceram. Soc.* **80**, [8] 1997-2004 (1997).
21. D. A. Litton and S. H. Garofalini. "Atomistic Structure of Sodium and Calcium Silicate Intergranular Films in Alumina by Molecular Dynamics". *J. Mat. Res.* **14** 1418-1429 (1999).
22. D. A. Litton and S. H. Garofalini. "Molecular Dynamics Simulations of Calcium Alumino-Silicate Intergranular Films on (0001) Alumina Faces". *J. Am. Ceram. Soc.* **83**, [9] 2273-2281 (2000).
23. C. M. Marian, M. Gastreich and J. D. Gale. "Empirical two-body potential for solid silicon nitride, boron nitride, and borosilazane modificaltion". *Phys. Rev. B* **62** 3117-3124 (2000).
24. M. Yoshiya, K. Tatsumi, I. Tanaka and H. Adachi. "Theoretical study on the chemistry of intergranular glassy film in Si3N4-SiO2 ceramics". *J. Am. Ceram. Soc.* **85**, [1] 109-112 (2002).
25. G. S. Painter, P. F. Becher and E. Y. Sun. "Bond energetics at intergranular interfaces in alumina-doped silicon nitride". *J. Am. Ceram. Soc.* **85**, [1] 65-67 (2002).
26. G. Pezzotti and G. S. Painter. "Mechanisms of dopant induced changes in intergranular SiO2 viscosity in polycrystalline silicon nitride". *J. Am. Ceram. Soc.* **85**, [1] 91-96 (2002).
27. G. S. Painter, P. F. Becher, H.-J. Kleebe and G. Pezzotti. "First-principles study of the effects of halogen dopants on the properties of intergranular films in silicon nitride ceramics". *Phys. Rev. B* **65** 064113-1-11 (2002).
28. W.-Y. Ching, S.-D. Mo and Y. Chen. "Calculation of XANES/ELNES spectra of all edges in Si3N4 and Si2N2O". *J. Am. Ceram. Soc.* **85**, [1] 11-15 (2002).

29. S. H. Garofalini and S. Zhang. in *Mat. Res. Soc. Symp. Proc.* 191-200 (Matls. Res. Soc., Pittsburgh, PA., Boston, 2003).
30. S. H. Garofalini and W. Luo. "Molecular dynamics simulations of calcium silicate intergranular films between silicon nitride crystals". *J. Am. Ceram. Soc.* **86**, [10] 1741-1752 (2003).
31. X. Su and S. H. Garofalini. "Role of Nitrogen on the Atomistic Structure of the Intergranular Film in Silicon Nitride: A Molecular Dynamics Study". *J. Mat. Res.* **19** 3671-3678 (2004).
32. X. Su and S. H. Garofalini. "Atomistic structure of calcium silicate intergranular films between prism and basal planes in silicon nitride: a molecular dynamics study". *J. Mat. Res.* **19** 752-758 (2004).
33. P. Rulis, et al. "Electronic structure and bonding of intergranular glassy films (IGF) in polycrystalline Si3N4: Ab-initio studies and classical MD simulations". *Phys.Rev. B* **71** 235317-1-10 (2005).
34. S. Zhang and S. H. Garofalini. "Molecular dynamic computer simulations of the interface structure of calcium-alumino-silicate intergranular films between combined basal and prism planes of a-Al2O3". *J. Am. Ceram. Soc.* **88**, [1] 202-209 (2005).
35. S. Blonski and S. H. Garofalini. "Molecular Dynamics Simulations of α-alumina and γ-alumina Surfaces". *Surf. Sci.* **295** 263-274 (1993).
36. J.-M. Delaye, V. Louis-Achille and D. Ghaleb. "Modeling oxide glass with Born-Mayer-Huggins potentials: Effect of composition on structural changes". *J. Non-Cryst. Sol.* **210** 232-242 (1997).
37. B. P. Feuston and S. H. Garofalini. "Empirical Three-Body Potential for Vitreous Silica". *J. Chem. Phys.* **89**, [9] 5818-5824 (1988).
38. D. A. Litton and S. H. Garofalini. "Vitreous Silica Bulk and Surface Self-Diffusion Analysis by Molecular Dynamics". *J. Non-Cryst. Sol.* **217** 250-263 (1997).
39. M. Doblinger, D. J. H. Cockayne, R. R. Meyer, A. I. Kirkland and D. Nguyen-Mahn. in *13th European Microscopy Congress* 59-63 (2004).
40. S. Zhang and S. H. Garofalini. "Molecular dynamics simulations of the effect of the composition of calcium alumino-silicate intergranular films on alumina grain growth". *J. Phys. Chem. B* (in press).

Table I. Parameters for Modified BMH Pair Potential

Atom Pair	A_{ij} (fJ)	β_{ij} (pm)	ρ_{ij} (pm)
O-O	0.0725	234	29
Si-Si	0.1877	230	29
Al-Al	0.0500	235	29
Ca-Ca	0.7000	230	29
Si-Al	0.2523	233	29
Si-Ca	0.2215	230	29
Al-Ca	0.2178	230	29
Si-O	0.2962	234	29
Al-O	0.2490	234	29
Ca-O	0.5700	234	29
N-N	0.07241	220	29
Si-N	0.7500	220	26.1
Ca-N	0.8600	223	29
N-O	0.1350	221	29

Table II. Parameters for 3-body Potential

Atom triplet	λ_{ij} (fJ)	γ_{ij} (pm)	r_{ij}^{0} (pm)	θ_{jik}^{0}
Al/Si/Ca-O-Al/Si/Ca	0.001	200	260	109.5
O-Al/Si-O	0.024	280	300	109.5
Si-N-Si	0.035	260	280	120
O/N-Si-O/N	0.024	280	300	109.5

Table III. Composition and system sizes for the alumina systems.

Sample No.	Mole ratio $CaO : Al_2O_3 : SiO_2$	Specific Number of Atoms				TOTAL
		Ca	Al	Si	O	
112	1:1:2	665	1330	1330	5320	8645
313	3:1:3	1296	864	1296	5184	8640
216	2:1:6	640	640	1920	5440	8640
14535	14:5:35	770	550	1925	5445	8690
334	3:3:4	720	1440	960	4800	7920
123	1:2:3	400	1600	1200	5200	8400
221	2:2:1	1000	2000	500	5000	8500
121	1:2:1	550	2200	550	4950	8250
141	1:4:1	340	2720	340	5100	8500

1. The number of atoms in the alumina crystals in the simulation are 7920 and 8640 for basal and prism orientations, respectively.

Table IV. Composition and system sizes for the nitride systems.

IGF composition	IGF atom numbers	Crystal surfaces
88% SiO_2 12% CaO N/(N+O)=0%	Si: 1048; O: 2238; Ca: 142 Total: 3428	Basal-prism combination
88% SiO_2 12% CaO N/(N+O)=15%	Si: 1097; O: 1858; N: 324; Ca: 150 Total: 3429	Basal-prism combination
88% SiO_2 12% CaO N/(N+O)=30%	Si: 1146; O: 1487; N: 640; Ca: 155 Total: 3428	basal-prism combination
88% SiO_2 12% CaO N/(N+O)=50%	Si: 1205; O: 1032; N: 1028; Ca: 164 Total: 3429	basal-prism combination

The nitride crystals total 14112 atoms.

Table V. Heat and quench process

Quench temperature	Conditions	Duration Nitride	Duration Alumina
10000K	NVE, crystal frozen	10 ps	10 ps
8000K	x, y constant, P_z=5 GPa	10 ps	20 ps
6000K	As above	10 ps	40 ps
4000K	As above	40 ps	40 ps
3000K	x, y constant, P_z=3 GPa, crystal de-frozen	40 ps	50 ps
2000K	NpT, p=0.1 MPa	10 ps	100 ps
1000K	As above	10 ps	100 ps
300K	As above	20 ps	20 ps

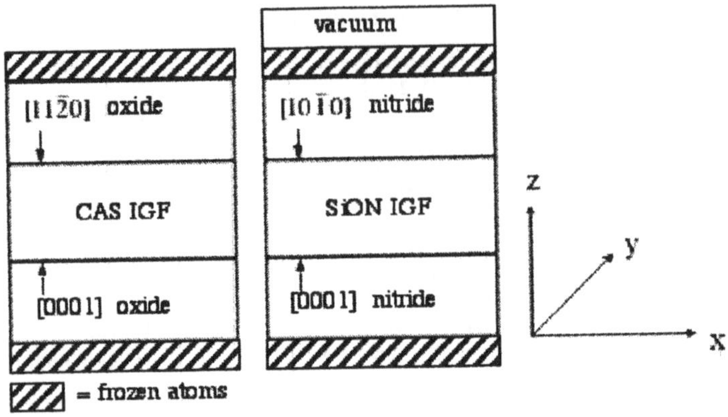

Fig 1. Schematic diagram of the crystals (oxide, nitride) separated by the IGF.

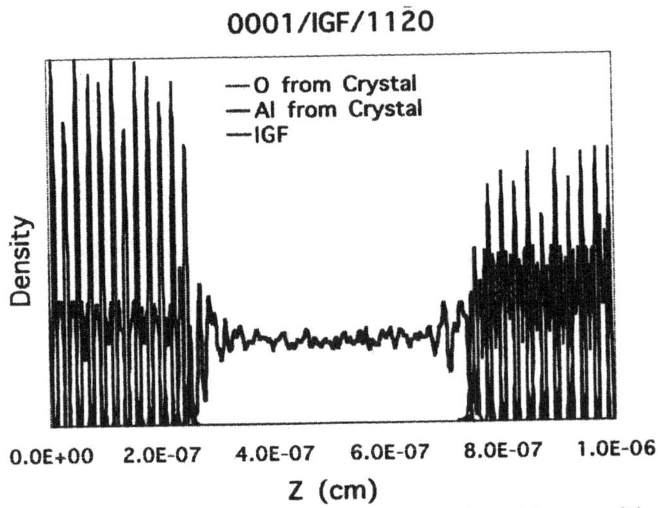

Fig 2. Density profile of the atoms in the alumina crystals and the sum of the atoms in the 112 IGF as a function of distance in Z, with the basal oriented crystal on the left and the prism oriented crystal on the right. Density oscillations seen in the IGF near the interfaces

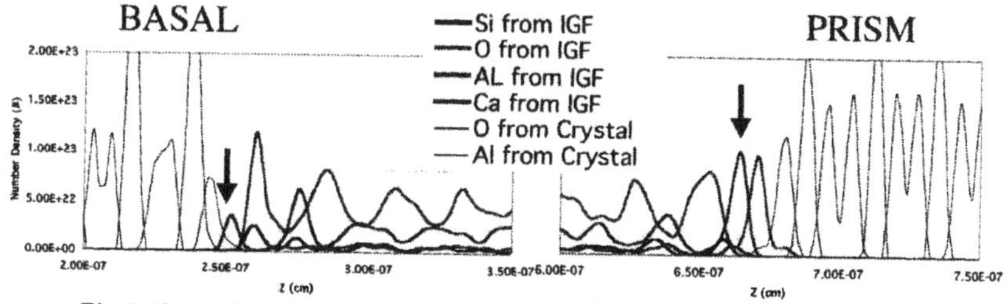

Fig 3. Expanded view of the density profiles of individual species in the 112 IGF and crystals in the alumina system. Arrows point to the first adsorbed Al peak on each side.

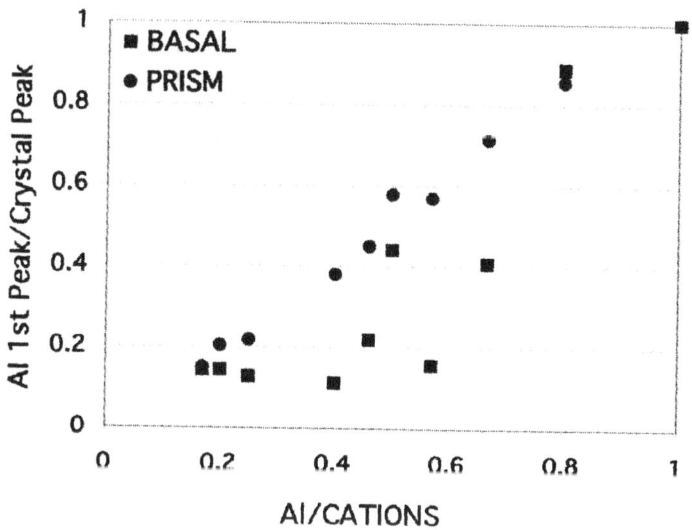

Fig 4. Normalized first adsorbed Al peak at the basal and prism surfaces as a function of Al concentration showing more rapid adsorption on the prism side, indicative of anisotropic growth over much of these compositions.

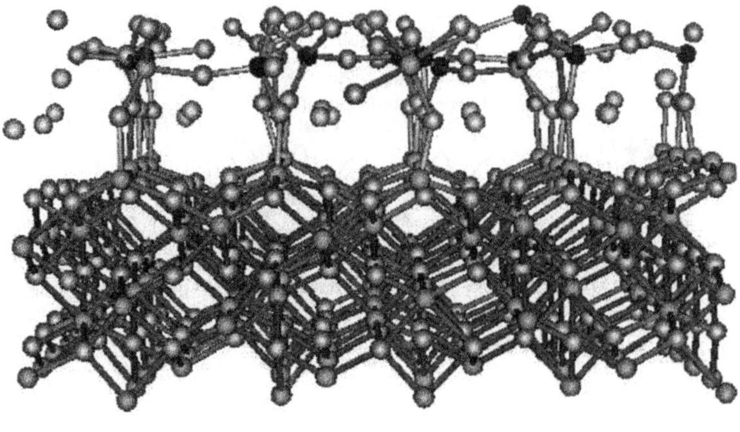

Fig 5 Sideview of atoms in the basal oriented crystal interfaced with atoms from the IGF. Cage structure forms at the interface, allowing yellow Ca ions siteing. O=grey, Al=red, Si=blue, and Ca=yellow. Ca-O bonds not drawn for clarity.

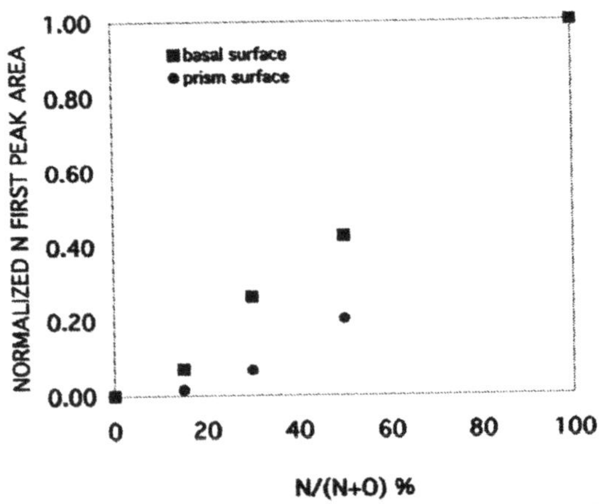

Fig 6. Normalized first adsorbed N peak as a function of N concentration in the nitride system showing more rapid adsorption onto the basal surface than the prism surface.

MODELING OF METAL/CERAMICS WETTING SYSTEMS

Takehiko Makino and Shun-ichiro Tanaka
Department of Engineering Physics, Electronics and Mechanics
Nagoya Institute of Technology, Gokiso-cho, Showa-ku, Nagoya, 466-8555, Japan

ABSTRACT

Wetting can be regarded as a kind of effective nanostructure-forming process. To control the structure, a study on the relationship between atomic interactions and the resultant wetting behaviors is required. To model the wetting system, two sets of interatomic potentials for Metal/MgO(100) systems are derived from first principles calculation results for the simple configurations. A molecular dynamics method is applied to simulate the system and shows that Al atoms wet better than Sn atoms on the MgO(100) substrate. This tendency is consistent with the experimental results of contact angles.

INTRODUCTION

Nanoscale wetting behavior has been an important subject particularly for the study of quantum dots[1], functional films[2] and surfaces[3]. Modeling a target system by setting an interatomic potential and investigating the behavior of the modeled system by molecular dynamics (MD) calculation is an effective method of exploring nanostructures. However, wetting phenomena are difficult to predict since these systems consist of three components, namely, the solid surface, liquid surface and liquid/solid interface. The subtle energy balance of these components determines the wettability and the interfacial structure formation. Considering the uniqueness of the wetting systems, it is strongly desirable to construct interatomic potentials following a consistent policy.

In our previous study[4], the interatomic potentials were derived from the first principles molecular orbital calculations and the difference in wetting behavior caused by surface orientation (Cu(111) and Cu(100)) was reported. The difference was interpreted by the relative depth of the interatomic potentials of each component. Extension to other substrate systems, namely semiconducting or insulating materials, is necessary to explore the possibility of the nanoscale world.

Ionic systems such as the oxides have been treated classically by rigid ion model[5], shell model[6], compressible ion model[7] and variable charge model[8, 9]. The development undergoes the variety of the treatment of the electronic polarizability[10] and charge transfer. The transferability has been also pursued. However, if the target is limited to a simple wetting system, of which substrate has non-polar surface, and the total energies can be obtained for any given configuration of atoms, it is possible to derive the interatomic potentials for the specific system without introducing above models.

There are a few MD studies[11, 12] on wetting dynamics of metal/oxide, that is Al/Al_2O_3, although the adsorbate and the substrate constituting metal are identical. The energetics of wetting of TiC and TiN was treated using first principles calculations[13]. For some Metal/MgO(100) interfaces, interatomic potentials have been constructed from first principles results[14]. Pd/MgO(100) system has treated by constructing the potentials derived from first principles results and the equilibrium structure of Pd cluster on the substrate is simulated by MD calculations[15].

As the first step of our approach, we selected MgO(100) as the substrate. Experimentally, Metal/MgO systems have been well studied[16, 17] and the wetting behaviors are classified as reactive and non-reactive[18]. It is also summarized[10] that the deposition behaviors, known as Stranski-Krastanov and Volmer-Weber growth mode, depend on adsorbates, surface directions and experimental techniques of surface preparation.

In the present study, we model the interatomic potentials from the first principles calculations and confirm the applicability of the obtained potentials to treat the dynamical wetting behavior of Metal/MgO(100) systems.

PROCEDURES

Modeling of metal/ceramics systems

Metal/MgO(100) systems were selected because of the simplicity. The substrate surface is chosen since the non-polar surface hardly reconstructs[10]. With respect to the contrasting experimental results of reactive and non-reactive behavior[18], Al and Sn, respectively, are chosen as the liquid elements to be modeled.

Total energy calculations for substrate surface

At first, the substrate surface has to be treated properly. In the present study, all first principles calculations were executed by ABINIT code[19, 20]. This density functional calculation code employs pseudopotential plane-wave technique. We used Troullier-Martins type pseudopotentials[20] of local density approximation. Plane-wave kinetic energy cut-off is set to be 60 Ryberg. To construct the system that contains surface, the slab model in which 3 slab layers with 4 equivalent vacuum layers is used. The supercell contains 12 atoms. The atomic coordinates of MgO(100) surface is shown in Fig.1 (a).

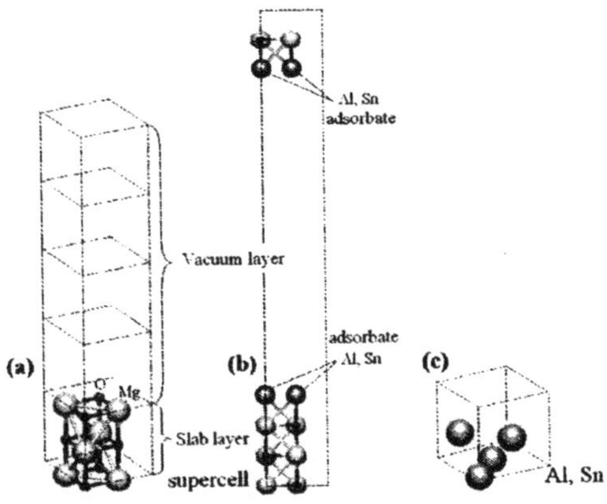

Figure 1. Atomic configurations of (a) substrate surface, (b) interface and (c) bulk (for liquid, see text) for the elements of wetting system. Dashed lines define the cell boundaries for each case.

Total energy change with isotropic expansion of the supercell (Fig. 1(a)) is shown in Fig.2. In spite of the existence of the surface, the lowest energy appears just at the bulk lattice parameter (0.4211nm). As the first step of our approach, we chose the substrate to be fixed. And therefore, long-range Coulomb interaction between Mg and O ions is not treated in the present study.

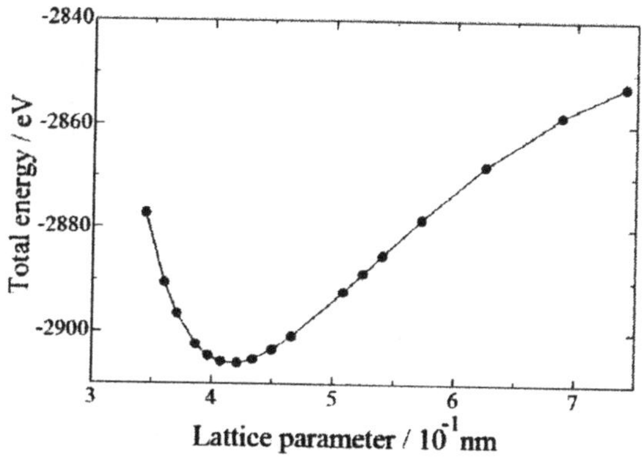

Figure 2. Total energy change with lattice parameter of MgO(100) substrate (with surface).

Total energy calculations and potential fitting for interface

The interatomic potential for the liquid/solid interface is the most important element for treating wetting systems. The atomic coordinates of the adsorbate/MgO(100) surface is shown in Fig.1 (b). For this case, 7 equivalent vacuum layers are used so as to displace the adsorbate . The calculation cell contains 16 substrate and 4 adsorbate atoms. To obtain interatomic potentials between the adsorbates and the substrate atoms, the total energies are calculated as the distance between the adsorbate and the substrate is changed. The interaction is considered up to third nearest atoms (9 atoms per an absorbate atom) on the surface. O-M (Metal adsobate, Al or Sn) and Mg-M interactions were derived as follows: Two energy curves were obtained by changing the distance between the substrate and M atoms just above (a) O site and (b) Mg site. To fit each curve, various combinations of O-M and Mg-M interactions are temporarily chosen. And finally, only one set of two interactions that can fit (a) and (b) curves simultaneously are determined under the condition that the O-M interaction is stronger than the Mg-M interaction. The parameter fitting was done by trial and error.

Total energy calculations and potential fitting for liquid

In our previous studies on wetting[4, 21] using molecular orbital calculations for non-periodical systems, liquid state consisting of more than 30 atoms was prepared by MD calculations using the empirical potential[22]. However, in the present study, we have to limit the number of atoms and apply periodic boundary condition. Therefore, we considered two alternatives, (a) fcc crystal structure with surface and (b) fcc bulk structure (with no surface). As the result of the fitting, the change in energy in (a) case was difficult to be fitted by simple two-body potential, as expected, because of the surface. Since the most important issue in the present study is the consistency for deriving the potentials in a system, we employed the condition (b), rather than use of another reported potential. The atomic coordinate of fcc structure employed is shown in Fig.1 (c). For Al and Sn cases, fcc structure is used to keep the same condition. The calculation cell contains 4 atoms. To obtain the interatomic potential between the liquid constituent atoms, the total energies as a function of the volume of the cell is calculated. The interaction is considered up to second nearest neighbors all through the volume change.

Potential function form

For comparison of the cause and results of the MD calculation, we employed simple interatomic functions. The potential functions are chosen to be Morse-type and Lennard-Jones-type two-body potential functions for the interface and the liquid, respectively. For the interface, the equation used is

$$\Phi_{\text{interface}}(r) = D_1 e^{-\lambda_1 (r - r_0)} - D_2 e^{-\lambda_2 (r - r_0)} \qquad (1)$$

where D (D_1 and D_2) , λ (λ_1 and λ_2), and r_0 are constants with dimensions of energy, reciprocal distance and distance, respectively, and r is the distance between two atoms. For the liquid, the equation used is

$$\Phi_{\text{liquid}}(r) = A\left\{(\sigma/r)^{p_1} - (\sigma/r)^{p_2}\right\} \qquad (2)$$

where A and σ are constants with dimensions of energy and distance, respectively, and p_1 and p_2 are dimensionless numbers. r is the distance between two atoms.

MD calculations

Using the derived potentials for the interface and the liquid, MD calculations are performed. For the calculations, numerical integration was executed by means of the velocity-Verlet method. The time step was 2 fs. Temperature control was carried out by the velocity scaling method applied for each 200 steps. The temperature was set to be 1050K. Periodic boundary conditions were applied to x-y directions (parallel to the substrate surface). As described before, the substrate atoms were fixed. The sizes of the substrates (number of units in each direction) were $14_{[100]} \times 14_{[010]} \times 1_{[001]}$. The liquid contained 256 atoms. The total number of atoms in a system was 1040. Initial configurations for the MD calculation simulated a typical sessile drop test by putting a cube-shaped liquid on the solid substrate surface.

RESULTS

Calculated total energies

Figure 3 shows the total energy differences of the Al/MgO(100) system. Solid circles represent the change in the energy with the distance between the Al atoms that are on the surface O atoms and the substrate. On the other hand, solid triangles are the results of the case that the Al atoms are on the surface Mg atoms. Compared to the solid triangles, the solid circles locate lower in energy and the position of its lowest point is at shorter distance.

Solid and dashed lines are drawn by curve fitting including the interactions between nearest 9 atoms on the surface. To fit to the two series of calculated points simultaneously, certain degree of deviation from calculated points was inevitable.

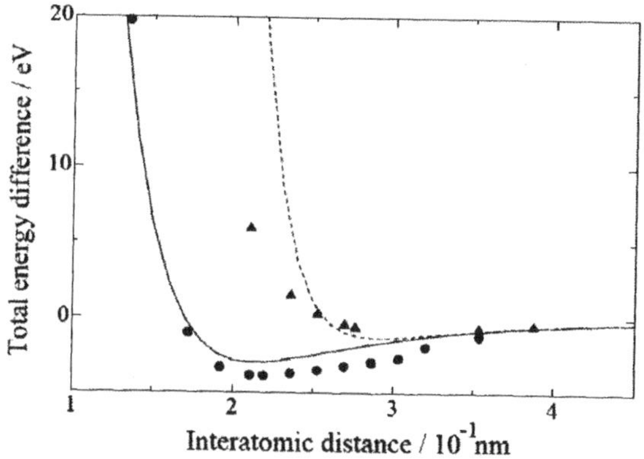

Figure 3. Change in total energy with the distance between the adsorbate and the substrate in Al/MgO(100) system. (a) MgO(O site)-Al interface (solid circle), (b) MgO(Mg site)-Al interface (solid triangle). Solid and dashed lines are drawn by curve fitting including the interactions between nearest 9 atoms on the surface.

Figure 4 shows the case for Sn/MgO(100) system. The relative position of the solid circles and triangles in this figure is closer than that in Fig. 3. The depth of the energies is slightly larger for Al/MgO system than for Sn/MgO system. The distance at which the energy is the lowest is shorter for Al/MgO than for Sn/MgO.

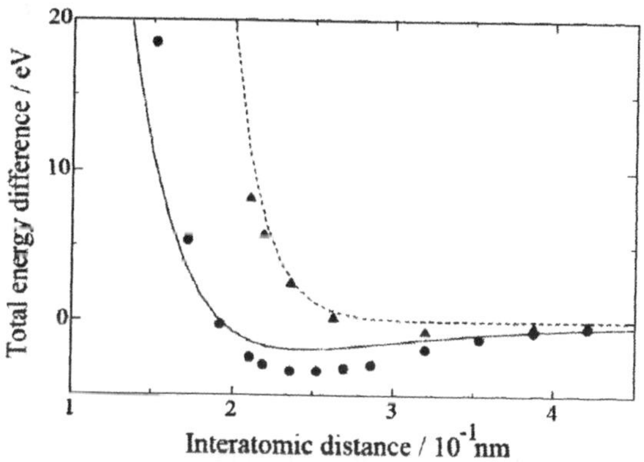

Figure 4. Change in total energy with the distance between the adsorbate and the substrate in Sn/MgO(100) system. (a) MgO(O site)-Sn interface (solid circle), (b) MgO(Mg site)-Sn interface (solid triangle). Solid and dashed lines are the results of fitting.

Figure 5 shows the total energy differences of (a) fcc Al bulk and (b) fcc Sn bulk. For Al, as expected, the lowest energy appears at the bulk lattice parameter (0.4049nm). The curve of Sn has deeper minimum.

Solid and dashed lines are drawn by curve fitting including the interactions up to second nearest neighbors in the bulk. In this case, the fitting is easier than the case of the interface as shown in Fig. 5.

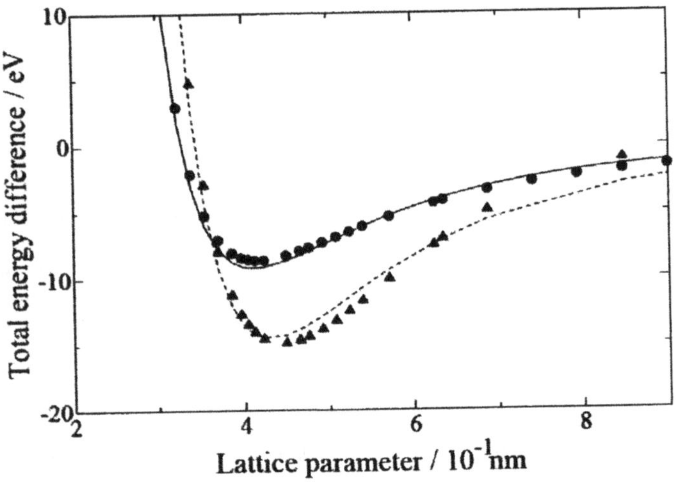

Figure 5. Change in total energy with the lattice parameter of (a) fcc Al bulk (solid circle) and (b) fcc Sn bulk (solid triangle). Solid line is drawn by including the interactions up to second nearest neighbors in the bulk. Dashed line is drawn in the same way.

Fitted potential energy functions

Figure 6 shows the fitted potentials for Al/MgO(100) system ((a) O-Al, (b) Mg-Al and (c) Al-Al interaction). The shapes of the potentials are different from each other. The O-Al potential has relatively deep minimum. On the other hand, the Mg-Al potential has only repulsive part. The Al-Al potential indicates their intermediate shape.

Figure 7 shows the potentials for Sn/MgO(100) system ((a) O-Sn, (b) Mg-Sn and (c) Sn-Sn interaction). Compared to the three potentials of Al system, the combination of three potentials in Fig. 7 is similar, however, the O-Sn potential is shallower and the Sn-Sn potential has deeper minimum. Accordingly, the relative depth of the Sn-Sn and O-Sn is closer.

Table 1 lists two sets of parameters for potential energy functions. For the interfacial interactions, the functions are formed by 5 parameters (listed above), and for the interaction of liquid atoms, the functions have 4 parameters (listed below).

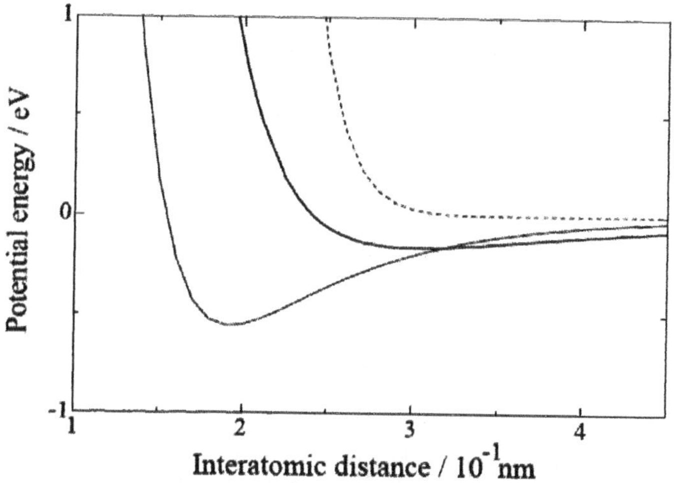

Figure 6. Potential functions for Al/MgO(100) system derived by fitting as shown in Fig. 3 for (a) O-Al (solid line) and (b) Mg-Al (dashed line) interactions. (c) Al-Al (heavy line) interaction is as in Fig. 5.

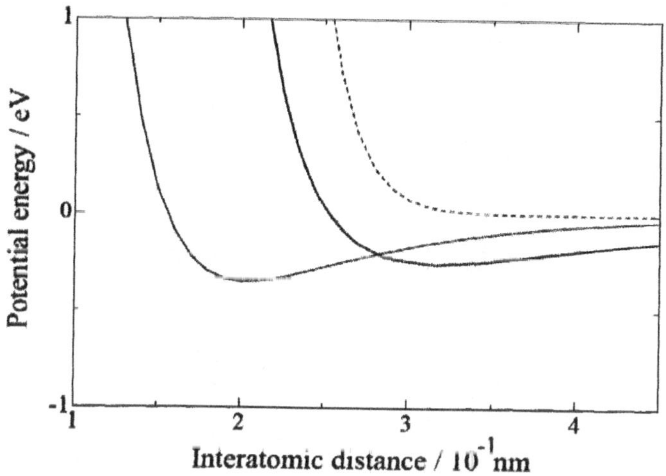

Figure 7. Potential functions derived by fitting as shown in Fig. 4 for (a) O-Sn (solid line) and (b) Mg-Sn (dashed line) interactions. (c) Sn-Sn (heavy line) interaction is as in Fig. 5.

Table 1. Parameters for interface (upper) and liquid (lower).

	O-Al	Mg-Al	Al-Al	O-Sn	Mg-Sn	Sn-Sn
D_1	19.00	20.00		18.00	20.00	
D_2	2.65	6.00		2.80	6.00	
λ_1	4.55	5.95		3.60	5.40	
λ_2	1.25	5.30		1.07	5.20	
r_0	0.94	2.05		0.70	2.06	
A			4.90			3.00
σ			2.41			2.53
p_1			4.61			5.00
p_2			4.21			3.95

MD calculations

Using the interatomic potentials obtained as shown in Figs. 6 and 7, MD calculations of two wetting systems were performed. It is confirmed that the derived potentials readily applicable to the MD calculations. Figure 8 shows the snapshots of the MD calculation results of (a) Al/MgO and (b) Sn/MgO systems after various periods at 1050 K. It is noticed that the Al atoms wet better than Sn atoms. Sn atoms keep their hemispherical shape even after 320ps.

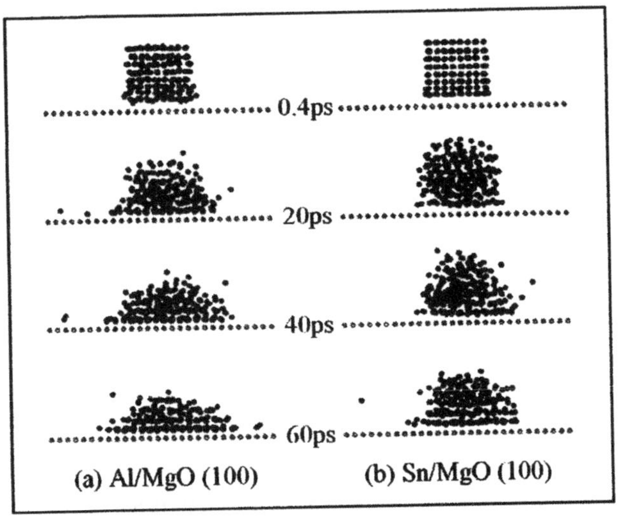

Figure 8. MD snapshots of (a) Al/MgO and (b) Sn/MgO systems. Dark gray circles represent liquid atoms and light gray circles represent substrates.

DISCUSSION

The potentials for the Al/MgO system derived in the present study is almost consistent with that reported by Yao and Zhang[14]. It is reported that the reaction in Al/MgO system starts at later stage of wetting[17]. Therefore, within the simulation time range, the present approach, in which the reaction phenomenon cannot be treated, will be justified.

The wetting behavior is understood by comparing the set of potential energies summarized in Figs. 6 and 7. The combination of relatively strong O-Al interaction and weak (shallow) Al-Al interaction results in wetting. Correspondingly, the combination of shallow O-Sn interaction and relatively strong Sn-Sn interaction results in non-spreading. The tendency is consistent with the experimentally measured contact angles (Al/MgO(100): 110 degree[17], Sn/MgO(100): 140 degree[16]).

Also from the combination of cohesive O-M (Al or Si) and repulsive Mg-M potentials, the configuration of the adsorbate at the surface can be deduced. Top view of the substrate surface shows that the Al atoms fall exactly on the O atoms as in Fig. 9. A few exceptional atoms that are out of alignment are away from the substrate.

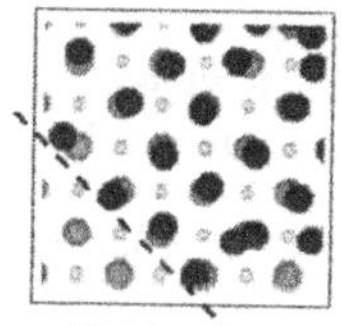

Figure 9. Top view of the configuration of Al atoms at wetting front on MgO(100) substrate. Dark gray circles are Al atoms. Light gray circles and small dots are O and Mg atoms, respectively.

The present study positively assumes that the substrate is fixed and the charge transfer is not considered. However, the potential for liquid is too simple. For determining the equilibrium shape of Pd cluster on the substrate, Vervish et al. [15] employed a second-moment approximation many-body potential. To develop the present approach without loss of simplicity, at least, we have to improve the potential for liquid even within the use of two-body functions.

CONCLUSION

1. Two sets of interatomic potentials for Metal/MgO(100) wetting systems are derived from the first principles calculation results of the simple configuration of the constituent atoms.

2. Al atoms wet better than Sn atoms on MgO(100) substrates. The tendency is consistent with the experimental results of contact angles. The behavior can be rationalized to the combination of the potential energy curves within a system.

REFERENCES

[1] J. E. Prieto and I. Markov, "Self-assembly of quantum dots: Effect of neighbor islands on the wetting in coherent Stranski-Krastanov growth", *Phys. Rev.* **B 69**, 193307 (2004).

[2] B. Ong, Y. Wu, L. Jiang, P. Liu, and K. Murti, "Polythiophene-based field-effect transistors with enhanced air stability ", *Synthetic Metals* **142**, 49-52 (2004).

[3] G. H. Gilmer, H. Huang, and C. Roland, "Thin film deposition: fundamentals and modeling ", *Comp. Mater. Sci.* **12**, 354-80 (1998).

[4] T. Makino, A. Kubo, H. Iida and S.-I. Tanaka," Interatomic potentials for metal/metal wetting systems", *Materials Science Forum* **502**, 75 (2005).

[5] M. Born, K. Huang, "Dynamical theory of crystal lattice", Clarendon Press, Oxford (1954).

[6] B. G. Dick, A. W. Overhauser, "Theory of the dielectric constants of alkali halide crystals", *Phys. Rev.* **112**, 90-103 (1958).

[7] N. A. Markes, M. W. Finnis, J. H. Harding, N. C. Pyper, "A physically transparent and transferable compressible ion model for oxides", *J. Chem. Phys.* **114**, 4406-14 (2001).

[8] F. H. Streitz, J. W. Mintmire, "Electrostatic potentials for metal-oxide surfaces and interfaces", *Phys. Rev.* **B 50**, 11996 (1994).

[9] S. Ogata, H. Iyetomi, K. Tsuruta, F. Shimojo, R. K. Kalia, A. Nakano, and P. Vashishta, "Variable-charge interatomic potentials for molecular-dynamics simulations of TiO2", *J. Appl. Phys.* **86**, 3036-41 (1999).

[10] V. E. Henrich, P. A. Cox and Cox, "*The surface science of metal oxides*", Cambridge University Press, Cambridge, (1994).

[11] T. P. Swiler, R. E. Loehman, "Molecular dynamics simulations of reactive wetting in metal-ceramics systems", *Acta mater.* **48**, 4419–24 (2000).

[12] G. Aral, "*Parallel molecular dynamics simulations of dynamics of oxidation and reactive wetting in metal/ceramic systems*", Thesis, Luisiana State University, (2003).

[13] S. V. Dudiy and B. I. Lundqvist, "Wetting of TiC and TiN by metals", *Phys. Rev.* **B 69**, 125421 (2004).

[14] Y. Yao and Y. Zhang, "Ab initio pair potential as metal-ceramic interface", *Phys. Lett.* **A 256**, 391-8 (1999).

[15] W. Vervisch, C. Mottet, J. Goniakowski, "Theoritical study of the atomic structure of Pd nanocluster deposited on a MgO(100) surface", *Phys. Rev.* **B 65**, 245411 (2002).

[16] K. Nogi, M. Tsujimoto, K. Ogino, N. Iwamoto, "Wettability of MgO single crystal by liquid pure Pb, Sn and Bi", Acta Metall. Mater. **40**, 1045-50 (1992).

[17] P. Shen, H. Fujii, T. Matsumoto, K. Nogi, "Wetting and reaction of MgO single crystals by molten Al", *Acta Mater.* **52**, 887-98 (2004).

[18] N. Eustathopoulos, M. G. Nicholas, B. Drevet, "*Wettability at High Temperatures*", Pergamon, Oxford (1999).

[19] X. Gonze, J.M. Beuken, R. Caracas, F. Detraux, M. Fuchs, G. M. Rignanese, L. Sindic, M. Verstraete, G. Zerah, F. Jollet, M. Torrent, A. Roy, M. Mikami, Ph. Ghosez, J. Y. Raty, D.C. Allan, "First-principles computation of material properties: the ABINIT software project", *Comp. Mater. Sci.* **25**, 478–92 (2002).

[20] http://www.abinit.org

[21] A. Kubo, T. Makino, D. Sugiyama and S.-I. Tanaka, "Molecular dynamics analysis of the wetting front structure in metal/metal systems", *Journal of Materials Science* **40**, 2395 (2005).

[22] L. A. Girifalco and V. G. Weizer, "Application of the Morse Potential Function to Cubic Metals", *Phys. Rev.* **114**, 687 (1959).

Wetting by Metal and Oxide Liquids

EVIDENCE OF A HIGH-TEMPERATURE WETTING TRANSITION BETWEEN Ag-CuO LIQUID ALLOYS AND YTTRIA-STABILIZED ZIRCONIA SUBSTRATES

K. S. Weil, J. T. Darsell, and J. Y. Kim
Pacific Northwest National Laboratory
P.O. Box 999
Richland, WA 99352

ABSTRACT

A series of sessile drop experiments was conducted at four different temperatures to determine the contact angles that form between Ag-CuO alloys and yttria-stabilized zirconia. When plotted as a function of composition, it was found that discontinuities exist in the resulting data at compositions that correspond to the liquid miscibility boundary displayed by the Ag-CuO system. Microscopic analysis conducted on quenched specimens confirms the first order transition and provides evidence that complete wetting likely takes place within the entire liquid coexistence phase region, in a fashion similar to that previously reported in other liquid metal systems.

INTRODUCTION

It is well known that immiscible liquids undergo a wetting transition, from partial to complete wetting, at a temperature below the critical point. This was originally predicted in 1977 by Cahn[1] and by Ebner and Saam.[2] Since that time, the phenomenon has gained a significant amount of theoretical attention, which has been surveyed in several excellent review papers.[3-5] Essentially, when two fluid phases coexist near their mutual critical point, one of the fluid phases will form a layer which intrudes between the other fluid phase and any third phase that happens to be present. In the case of a solid substrate adjacent two liquids present under near critical point conditions, one liquid will perfectly wet the solid and thereby encapsulate the second liquid phase excluding it from contact with the solid. As the two liquid system is moved away from its thermodynamic critical point, T_c, to a lower temperature, T_w, a phase transition will occur such that the intruding layer vanishes.

Over the past two decades, substantial progress has been made in the experimental study of this phenomenon with significant attention paid to the transitions observable in room-temperature binary liquids such as methanol-cyclohexane.[6,7] More recently, work by Freyland and Nattland[8] and by Chatain and Wynblatt and co-workers[9,10] provide experimental evidence of a similar phenomenon in liquid metal systems, namely Ga-Tl and Ga-Pb. Unlike the organic binary liquids, in which bonding is typically characterized by long-range van der Waals interactions, the wetting behavior observed in the metallic systems appeared to be indicative of short ranged interactions. For example in the Ga-Pb system, the wetting transition temperature was found to lie in the metastable region of the liquid miscibility gap (i.e. below the monotectic temperature), far from the critical temperature for the system at approximately $0.3T_c$.[10] In contrast, T_w in the methanol-cyclohexane lies above $0.5T_c$.[6]

In our recent development of braze filler metals based on the Ag-CuO system, we observed evidence of a first-order wetting transition. Like the previously studied liquid metal systems, the Ag-CuO phase diagram displays a wide region of liquid-liquid immiscibility, in this case between a silver-rich liquid L_2 and a copper oxide-rich liquid L_1, that lies above the 969°C monotectic reaction temperature[11] as shown in Figure 1. Rule of mixtures estimates of the densities of the two liquids in the miscibility gap indicate that the density of silver-rich L_2 should be greater than that of L_1 by nearly 40%. However as will be shown, due to its lower surface energy with the yttria-stabilized zirconia (YSZ) substrate the copper oxide-rich liquid maintains a condition of complete wetting despite being the lighter of the two fluids.

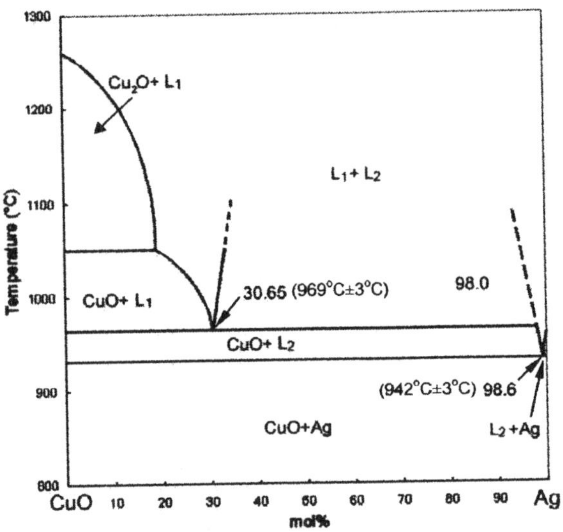

Figure 1 Ag-CuO phase diagram, from Ref [11] with updated invariant reaction temperatures based on Ref [12].

EXPERIMENTAL

Materials

YSZ discs measuring ~3 cm in diameter and ~0.5 cm in thickness were used in the wetting experiments. These were prepared by isostatically cold-pressing 8 mol% yttria-stabilized zirconia powder (YSZ, Tosoh) into pellets under a pressure of 150 MPa. These pellets were subjected to sintering in static air at 1450°C for 1 h. The sintered discs were then polished along one face to a 10 μm finish, ultrasonically cleaned in methanol, and dried at 300°C for 1hr prior to use. Polycrystalline YSZ cups containing 8 mol% yttria (Netzsch) and measuring 7 mm tall x 7 mm in diameter x 0.7mm wall thickness were used in the quenching experiments.

The Ag-CuO compositions employed in both the wetting and quenching experiments were prepared by mixing copper powder (99% purity; 1 – 1.5 μm average particle size; Alfa Aesar) and silver powder (99.9% purity; 0.5 – 1 μm average particle size; Alfa Aesar) in the appropriate ratios to yield the compositions given in Table I. It has been previously shown that the copper readily oxidizes to become CuO under the air brazing conditions.[12] The braze compositions given in Table 1 take this into account. The powder mixtures were then pressed into pellets measuring ~ 0.6 mm in diameter by 0.6 mm thick.

Testing and Characterization

Wetting experiments were conducted by the classical sessile drop technique.[13] Each Ag-CuO pellet was placed on a polished YSZ substrate and heated through a series of temperatures above which the pellet becomes molten. Changes in the braze droplet's base diameter, height, and angle of contact with the substrate were monitored by high speed video camera. The specimens were heated at 30°C/min to a sub-liquidus temperature of 900°C, at which point the heating rate was reduced to10°C/min for the rest of the cycle. Once stable at 900°C, the furnace temperature was raised to 950°C, where it remained for fifteen minutes. It was then increased again to a second isothermal hold at 1000°C for fifteen minutes, followed by a fifteen minute soak at 1050°C, and again at 1100°C. In this way, the shape of the sessile drop was allowed to stabilize for measurement at each temperature. Once the wetting experiment concluded, each sample was quenched at 10°C/min to 550°C, then furnace cooled to room temperature. Conversion of the analog video data to digital form for image analysis was performed using VideoStudio6™ (Ulead Systems, Inc.) frame capturing and video editing software. The digitalized images were subsequently imported into Canvas™ graphic software (version 8.0.5, Deneba Systems, Inc.) for droplet shape measurement. The outline of the droplet was circumscribed with a circle of best fit, which was used to determine the droplet's radius. The contact angle was then calculated from the radius and the diameter of contact between the droplet and the substrate.

Table 1. Target compositions of the Ag-CuO filler materials investigated in this study

Braze ID	Ag content (mol%)	CuO content (mol%)
CA0	100	0
CA01	99	1
CA01.4	98.6	1.4
CA01.8	98.2	1.8
CA02	98	2
CA03	97	3
CA04	96	4
CA05	95	5
CA07	93	7
CA08	92	8
CA10	90	10
CA16	84	16
CA34	66	34
CA69	30.7	69.3

A series of quenching experiments were conducted in an effort to capture the original molten state of the wetting speciemens. In each experiment a silver-copper oxide pellet was set into a YSZ crucible that was then placed onto a fecralloy harness as shown in Figure 2. The harness was tied to a fecralloy wire such that the specimen could be suspended in the hot zone of a vertical tube furnace. The specimen was heated at 3°C/min to the temperature of interest and held for 30 min to reach equilibrium. The insulation plug at the bottom of the furnace was then removed and the wire cut, allowing the specimen to drop into a room temperature water bath. Subsequent to quenching, the samples were sectioned and mounted for examination. Microstructural analysis was conducted on polished cross-sectioned samples by scanning electron microscopy (SEM; JEOL, JSM-5900LV) and energy dispersive X-ray analysis (EDX; Oxford Systems). The SEM is equipped with a windowless detector, allowing detection and quantitative measurement of both light and heavy elements. To avoid electrical charging of the samples in the SEM, they were carbon coated and grounded. Elemental profiles were recorded across joint interfaces in the line-scan mode.

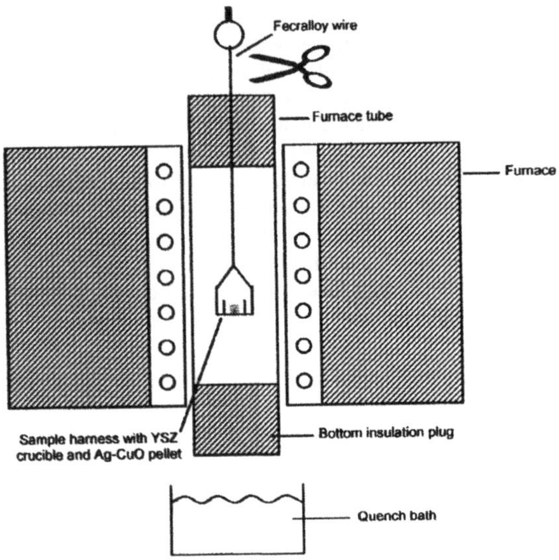

Figure 2 A schematic of the equipment used in the quenching experiments.

RESULTS AND DISCUSSION

Wetting Experiments

Shown in Figures 3(a) and (b) are the contact angles measured between molten Ag-CuO and YSZ as a function of copper oxide content and temperature. During the initial stage of heating each composition, liquefaction was observed to begin approximately at the temperature predicted by the Ag-CuO phase diagram,[5] implying that neither diffusion nor solid-state reaction kinetics significantly impeded the CuO and Ag from reaching equilibrium. In addition, the fifteen minute isothermal dwell times employed in the heating cycle appeared to be sufficiently long for the contact angles to reach their equilibrium values.

As seen in Figure 3(a), the addition of CuO causes a significant decrease in the contact angle between the molten drop and YSZ, from the non-wetting condition displayed by pure silver to an apparent minimum value reached at ~35mol% CuO. Much of the drop occurs in a narrow

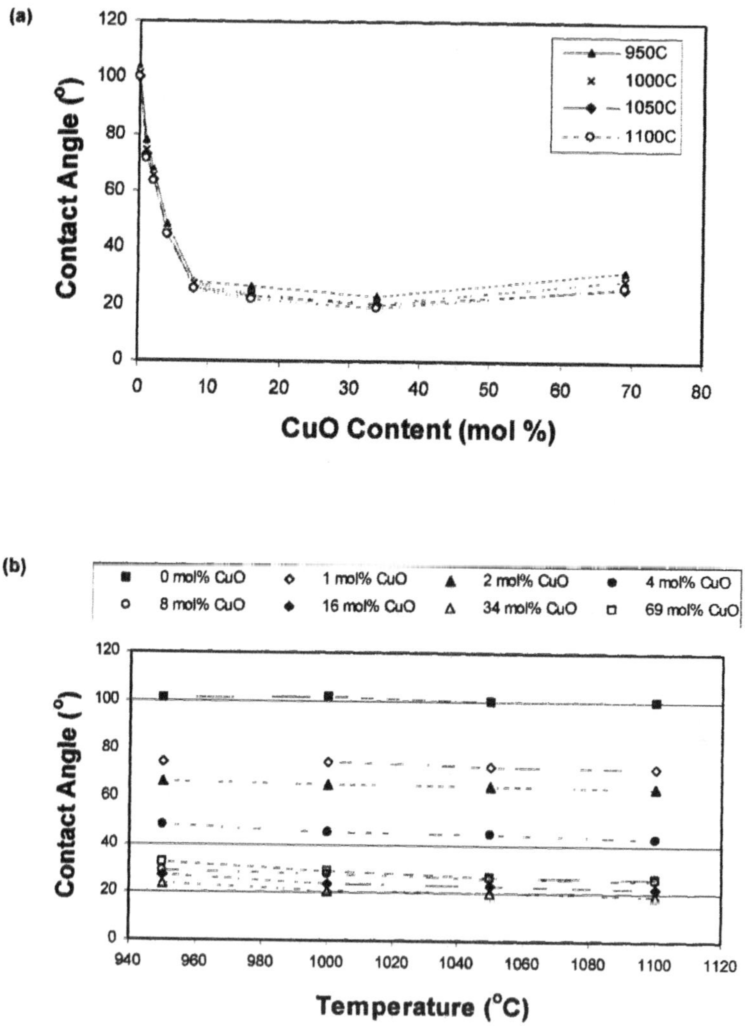

Figure 3 A plot of the contact angle formed between Ag-CuO and YSZ as a function of: (a) CuO content and (b) temperature.

compositional band bounded at an upper limit of $X_{CuO} \approx 4$mol%. In fact, each curve appears to consist of three distinct regions, as defined by obvious changes in slope: (1) $0 < X_{CuO} \lesssim 4$mol%, where the greatest decrease in contact angle occurs; (2) 4mol% $\lesssim X_{CuO} \lesssim$ 35mol%, denoted by a more gradual decrease in contact angle as a function of composition; and (3) 35mol% $\lesssim X_{CuO} \lesssim$ 69mol%, in which a small, but reproducible increase in contact angle is observed at the end point composition. In Figure 3(b), a second compositional effect on wetting behavior is observed. Below 4mol% CuO content (specimens CA0 – CA02), the contact angle appears to be nearly invariant of temperature. However at this threshold and above, the contact angle displays a small but measurable decrease with increasing temperature.

When the first derivative of the contact angle with respect to composition is plotted as a function of composition, as in Figure 4, a "near step" change can be observed in each isothermal curve [which appear to lie upon each other, but as shown in Figure 5(a) do not]. Note that changes in the slope of the isothermal curves in Figure 3(a) are not completely abrupt, leading to some ambiguity in identifying the exact composition at which each break takes place. In an effort to provide some clarity, a tangent line was drawn from the initial point on the derivative curve to the abscissa or compositional axis, by focusing on the narrow low-CuO concentration portion of the diagram as in Figure 5(a). Direct comparison with the Ag-CuO phase diagram of Nishiura et al.,[14] Figure 5(b), indicates that the discontinuities occur approximately at the corresponding silver-rich miscibility gap boundary compositions for each temperature. That is, the first order change in equilibrium contact angle appears to be due to the formation of a second liquid phase.

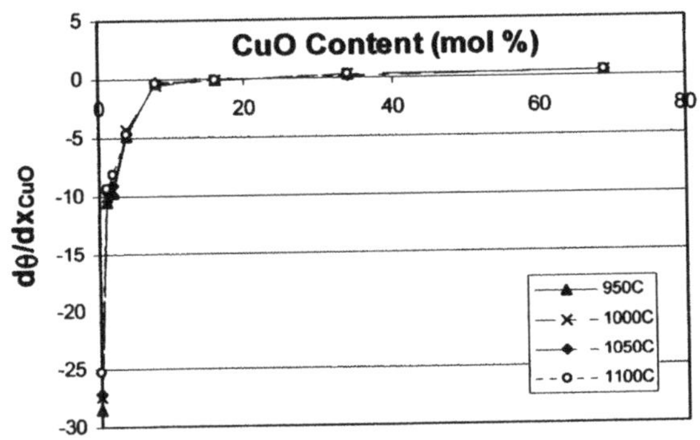

Figure 4 A plot of the derivative of contact angle as a function of composition for each hold temperature.

Quenching Experiments

Subsequent to the wetting experiments, two series of quenching experiments were conducted at 1000°C and 1100°C in an effort to capture direct evidence of the nature of the wetting transition observed in Figures 3 and 4. Specifically, pellets were prepared with the following compositions in order to bracket the silver-rich miscibility gap boundary compositions at each of the two temperatures: (a) 1000°C; 1mol% CuO, 1.4 mol% CuO, 1.8 mol% CuO, 3mol% CuO, 5 mol% CuO, 7 mol% CuO, and 10 mol% CuO and (b) 1100°C; 3 mol% CuO, 5 mol% CuO, 7 mol% CuO, and 10 mol% CuO, with the balance silver in each case. Shown in Figures 6 and 7 are back-scattered electron images of typical as-quenched monophasic and diphasic Ag-CuO liquids in contact with a YSZ crucible after allowing each sample to equilibrate at 1000°C and 1100°C, respectively.

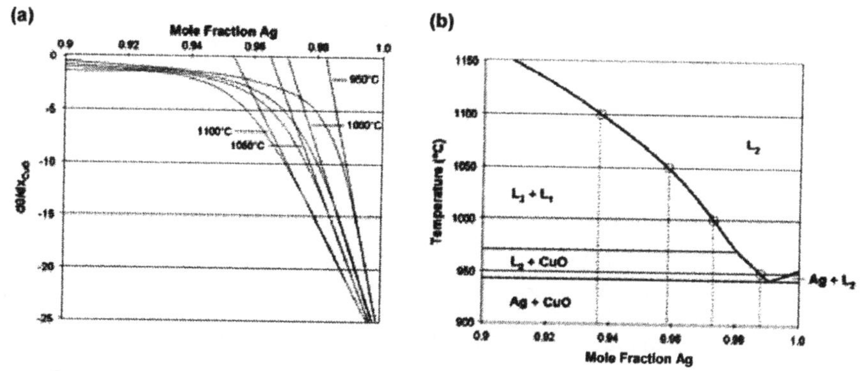

Figure 5 Plots of the silver-rich portions of (a) Figure 4 and (b) Figure 1. A comparison between the two indicates the correspondence between the discontinuities in the isothermal contact angle curves and the miscibility gap boundary compositions at each temperature.

In the specimen wetted by 1mol% CuO in Ag (CA01) and quenched from an equilibrium temperature of 1000°C, shown in Figure 6(a), the copper oxide is observed both in the solidified silver matrix that made up the bulk of the liquid sessile drop and along the interface with the YSZ as fine-scale precipitates measuring approximately 1 – 2μm in size. Along the interface with the YSZ, both silver and copper oxide are observed. Reference to the Ag-CuO phase diagram suggests that this microstructure results from a single phase liquid, L_2. Results obtained upon quenching from a two-phase liquid region are found in Figure 6(b), which displays the interface within a CA03 specimen that was equilibrated at 1000°C. A higher magnification image is shown in Figure 6(d). Note the very fine-scale, interwoven monotectic structure in contact with the YSZ interface. Since it is the copper oxide-rich L_1 liquid that undergoes monotectic decomposition, these two micrographs provide evidence that this liquid preferentially wets the YSZ substrate, even though it is the lighter of the two liquids (i.e. of much higher oxide content) present at 1000°C. With a composition that lies further in the diphasic liquid region, the CA05 specimen as quenched from 1000°C is shown in Figure 6(c). Again the oxide-rich L_1 liquid appears to have segregated to the YSZ interface forming an even thicker intruding layer when molten, over six times that observed in the CA03 specimen.

The second series of micrographs (Figure 7), taken of specimens that were quenched from 1100°C, display characteristics with respect to interfacial microstructure that are similar to the first set except that the corresponding compositions have shifted to higher CuO content. For example, the sample that displays the microstructure of a quenched monophasic liquid is CA03, Figure 7(a), whereas at lower temperature this same composition yielded a diphasic structure. Similarly, in Figure 7(b) the remnant of the intruding oxide-rich liquid layer in the CA05 specimen equilibrated at 1100°C is much thinner than that equilibrated at 1000°C. Even the intruding liquid of the CA07 specimen in Figure 7(c) is thinner, up to the leading edge of the sessile drop which is dominated by the copper oxide rich L_I liquid. Note that this liquid encapsulates the silver-rich liquid, keeping it from wetting the solid substrate.

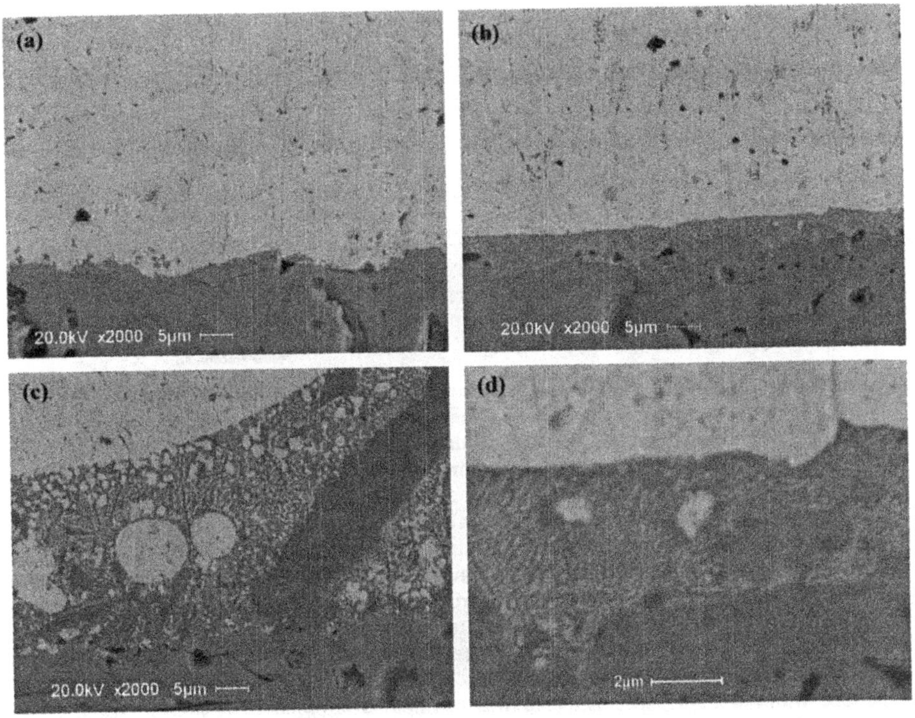

Figure 6 Cross-sectional SEM micrographs of the Ag-CuO/YSZ interface in the following specimens quenched from 1000°C: (a) CA01, (b) CA03, (c) CA05. A higher magnification micrograph of the interfacial region in the CA03 specimen is shown in (d). The YSZ appears as the darker gray, homogeneous material in the lower half of each picture, while silver has a lighter gray contrast. The copper oxide appears as the darker gray phase in the solidified material. The black regions are voids.

When compared with the contact angle data given in Figures 3 – 5, two major conclusions can be drawn from the microstructural data obtained on the quenched specimens: (1) the continuous and sharp decrease in contact angle observed between 0 and 8 mol% CuO in silver in Figure 3(a) is due to CuO adsorption at the silver-YSZ interface that leads to a decrease in interfacial energy and (2) the approximate first order change in contact angle with composition observed in Figure 5(a) is due to the formation of a diphasic liquid state as the miscibility gap

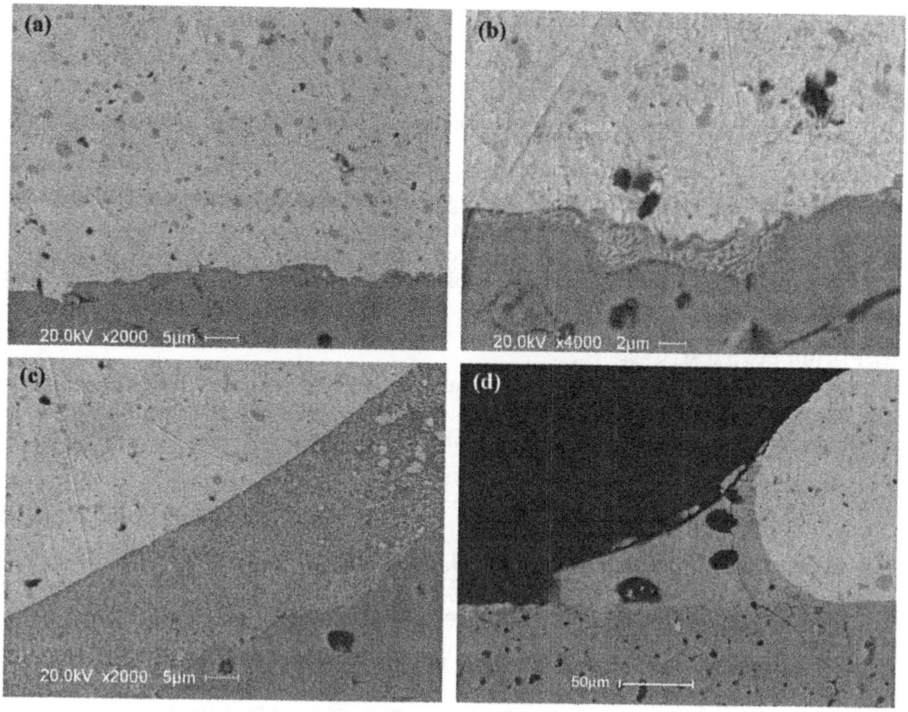

Figure 7 Cross-sectional SEM micrographs of the Ag-CuO/YSZ interface in the following specimens quenched from 1100°C: (a) CA03, (b) CA05, (c) CA07. A micrograph of the leading edge of the as-quenched CA07 sessile drop is shown in (d). The YSZ appears as the darker gray, homogeneous material in the lower half of each picture, while silver has a lighter gray contrast. The copper oxide appears as the darker gray phase in the solidified material. The black regions are voids.

boundary is isothermally crossed. Even though it is greater in density, the silver-rich liquid does not displace the copper oxide-rich liquid and is hindered from wetting the YSZ substrate by an intruding L_1 liquid film. A similar observation was made by Wang et al. of the wetting of graphite by Pb-Ni alloys.[15] In addition, evidence of encapsulation of the silver-rich liquid by the copper-rich liquid is a signature of complete wetting of L_2 by L_1. From the contact angle data in Figures 3(a) and 5(a), this phenomenon appears to occur at the monotectic temperature, i.e. the lowest temperature at which the two liquids can coexist in equilibrium, suggesting that if a critical wetting transition of the type predicted by Cahn exists, it must occur in the metastable region of liquid-liquid coexistence and therefore would only be observed upon careful super-cooling below the monotectic temperature. As has been previously noted by Chatain et al.[10] in their work on wetting transitions in liquid Ga-Pb alloys, this corresponds to the "wet-not-wet" scenario discussed by Pandit and Fisher.[16]

CONCLUSIONS

The addition of copper oxide to silver enhances wetting on YSZ substrates via CuO adsorption at the interface. As the miscibility gap is entered isothermally from the monophasic L_2 liquid region, the wetting mechanism changes abruptly. Two immiscible liquids form within the sessile drop and the one containing the higher copper oxide content intrudes between the second liquid and the substrate forming a thin liquid film. In addition, evidence of complete wetting of the two liquids down to the monotectic temperature for the Ag-CuO system implies that a critical wetting transition, if it exists, likely takes place in the metastable liquid-liquid domain.

ACKNOWLEDGMENTS

This work was supported by the U. S. Department of Energy as part of the SECA Program. The Pacific Northwest National Laboratory is operated by Battelle Memorial Institute for the United States Department of Energy (U.S. DOE) under Contract DE-AC06-76RLO 1830.

REFERENCES

1 J. W. Cahn, J. Chem. Phys., **66**, 3667 (1977).
2 C. Ebner and W. F. Saam, *Phys. Rev. Lett.*, **38**, 1486 (1977).
3 D. Bonn and D. Ross, *Rep. Prog. Phys.*,**64**, 1085 (2001).
4 P. G. de Gennes, *Rev. Mod. Phys.*, **57**, 827 (1985).
5 M. Schick, "Liquids at Interfaces," *Les Houches Sessions*, ed. J. Charvolin, 1990, Amsterdam: Elsevier, p. 415.
6 M. R. Moldover and J. W. Cahn, *Science*, **207,** 1073 (1980).
7 H. Kellay, D. Bonn, and J. Meunier, *Phys. Rev. Lett.*, **71**, 2607 (1993).
8 W. Freyland and D. Nattland, *Ber. Bunsenges. Phys. Chem.*, **102**, 1 (1998).
9 H. Shim, P. Wynblatt, D. Chatain, *Surf. Sci.*, **476**, L273 April (2001).
10 D. Chatain and P. Wynblatt, Surf. Sci., **345**, 85 (1996).
11 Z.B. Shao, K.R. Liu, L.Q. Liu, H.K. Liu, S. Dou, *J. Amer. Cer. Soc.*, **76**, 2663 (1993).
12 J. T. Darsell and K. S. Weil, *J. Ph. Equil. Diff.*, in press.
13 D.A. Mortimer and M.G. Nicholas, *J. Mater. Sci.*, **8**, 640 (1973).
14 H. Nishiura, R. O. Suzuki, K. Ono, and L. J. Gauckler, *J. Amer. Cer. Soc.*, **81**, 2181 (1998).
15 Z. Wang, P. Wynblatt, and D. Chatain, Interf. Sci., **7**, 173 (1999).
16 R. Pandit and M. E. Fisher, *Phys. Rev. Lett.*, **51**, 1772 (1983).

THERMOCHEMISTRY OF SAPPHIRE WETTING BY STEEL CONTAINING BOTH ALUMINUM AND TITANIUM IN NITROGEN ATMOSPHERES

P. Darrell Ownby and Benjamin T. Eldred
University of Missouri – Rolla
Department of Materials Science and Engineering
222 McNutt Hall
Rolla, MO 65401

ABSTRACT

A modified sessile drop technique was used to investigate the wetting of steels containing aluminum and/or titanium as a function of furnace atmosphere. It was found that the steel chemistry and furnace atmosphere had little effect on wetting except in the case of a particular ultra-low carbon steel containing both aluminum and titanium. This steel was found to show significantly lower contact angles than any other steel tested when it was in an atmosphere of pure hydrogen. As nitrogen was added to the atmosphere, the contact angle increased monotonically and irreversibly. The interaction between aluminum, titanium, and nitrogen is explained in terms of first-order interaction coefficients available in thermodynamic literature.

INTRODUCTION

Clogging of the submerged entry nozzle, a component that directs the flow of molten steel into the mold during continuous casting, has been a major problem in the steelmaking industry for many years. The American Iron and Steel Institute along with the Department of Energy sponsored a research program at the University of Missouri-Rolla to investigate the problem. One phase of this investigation was a sessile drop study of the wetting phenomena of steel on various refractory substrates as a function of steel chemistry, furnace atmosphere, and refractory composition. During the study it was found that steel containing additions of both aluminum and titanium displayed anomalous wetting behavior in the presence of nitrogen not reported in any previous publication. Specifically, this steel showed a monotonic, irreversible increase in contact angle as a function of nitrogen content in the furnace atmosphere.

EXPERIMENTAL PROCEDURE

The steels used throughout this study were prepared at UMR. Iron, with the appropriate additions of carbon, aluminum, and/or titanium, was melted in a 50-pound capacity induction furnace. A detailed discussion of the furnace practices and sampling procedure can be found in Ramachandran's thesis[1]. The steel compositions used are presented in Table I. Naming is based on carbon content (MC = medium carbon, LC = low carbon, ULC = ultra low carbon) and whether or not Al or Ti has been added to the melt.

Sessile drop studies were carried out in an atmosphere-controlled alumina tube furnace equipped with a doser tube, illustrated in Figure 1. The doser tube is a closed-on-one-end alumina tube with a small hole drilled near the tip of the closed end that allows a drop of steel to be placed on the substrate after it has equilibrated with the furnace atmosphere at the desired experimental temperature. The doser tube technique has been shown by Ownby and Li[2] to be a superior method of drop formation for samples which are highly sensitive to oxygen partial pressure. The drops produced by the doser averaged 1.5 grams and 6 to 9mm diameter at the

Table I – Compositions of experimental steels.

Label	%C	%Al	%Ti
MC-AlTi	0.289	0.12	0.10
LC (1)	0.02		
LC (2)	0.03		
ULC	0.0010		
ULC-Al	0.0030	0.1235	0.0006
ULC-Ti	0.003		0.10
ULC-AlTi	0.0024	0.1162	0.1169
ULC-AlAl	0.006	0.22	

In all cases, the following impurity limits apply:
Mn<0.04, Si<0.01, S<0.006, P<0.005, Cu<0.003, Ni<0.007, Cr<0.01, Mo<0.01

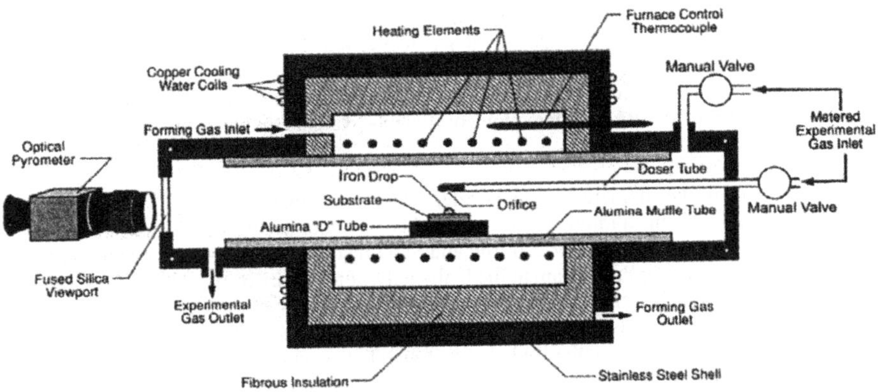

Figure 1 – Schematic Diagram of Sessile Drop Furnace.

base. Oxygen partial pressure in the furnace was monitored with a thoria oxygen sensor. A digital video camera was used to film each experiment, and the contact angle was measured from high resolution still images captured from the film. The error in contact angle measurement is estimated to be no larger than ±2°. The furnace atmosphere was regulated by two mass flow controllers flowing nitrogen and hydrogen.

Two types of sessile drop experiments were conducted. In the earliest experiments, the drop was simply held for eight hours in a constant atmosphere and the contact angle was measured as a function of time. The atmosphere was either pure hydrogen, forming gas

(90 vol% N_2 + 10 vol% H_2), or a combination of hydrogen and forming gas. During the course of the investigation, it was found that the contact angle of steel ULC-AlTi, ultra-low carbon with both aluminum and titanium additions, varied with the nitrogen content of the furnace atmosphere. A procedure was developed to measure the effect of varying atmospheric nitrogen in a single furnace run. A drop was formed in pure hydrogen, and then nitrogen was added in a stepwise manner, illustrated in Figure 2, allowing time for the drop to equilibrate at each nitrogen level. After a peak of about 17 vol% nitrogen, the nitrogen flow was decreased back to zero in the same stepwise manner.

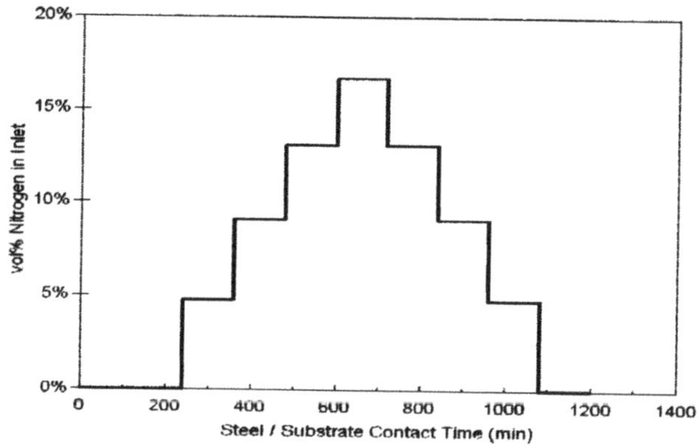

Figure 2 – Atmospheric nitrogen as a function of steel / substrate contact time.

RESULTS AND DISCUSSION

Contact Angle as a Function of Time

The first experiments run were exploratory – a drop of steel was formed in either forming gas (90 vol% N_2 + 10 vol% H_2) or hydrogen, and the contact angle was measured as a function of time. Pure electrolytic iron was also tested for comparison. The results are shown in Figure 3, except for steel ULC-AlTi in hydrogen which will be explained separately. All of the steel samples, with the exception noted, had a contact angle that started between 102°-115° and dropped to 100°-105° within a few hours. All of the steel samples showed a distinctly different contact angle from pure iron (10°-15° lower), but there were not significant differences between the various steel samples.

The most probable explanation for this behavior is the issue of concentration. Adding even a small amount of a surface active element to a pure system greatly increases the entropy of the interfacial region, causing a decrease in γ_{SL}, the solid-liquid interfacial energy. The contact angle, θ, is related to the solid-liquid, solid-vapor, and liquid-vapor interfacial energies by the Young Equation:

$$\cos\theta = \frac{\gamma_{SV} - \gamma_{SL}}{\gamma_{LV}} \qquad \text{...[1]}$$

If γ_{SV} and γ_{LV} remain constant, a decrease in γ_{SL} causes an increase in cosθ, which causes a decrease in θ for $90° < \theta < 180°$. In short, adding a surface active element to a pure, non-wetting system causes it to become more wetting. However, after the ≈10° drop in contact angle caused

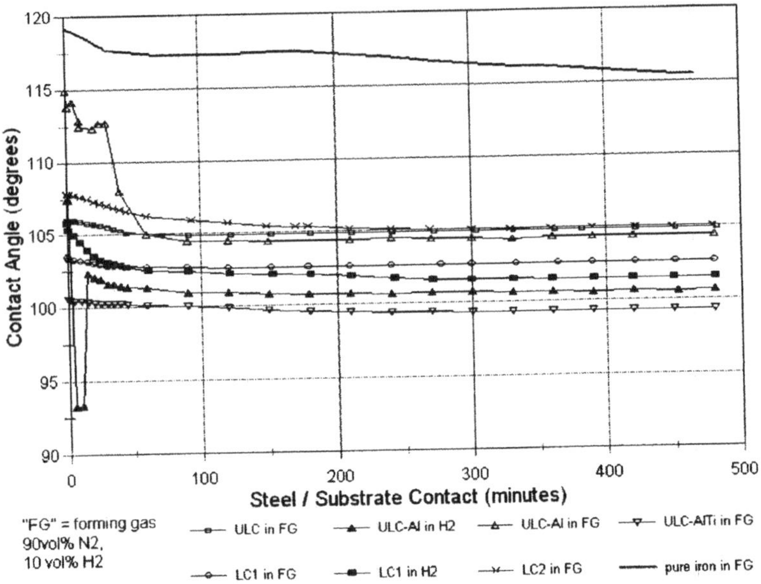

Figure 3 – Contact angles of various steels on sapphire at 1600°C. The case of ULC-AlTi in hydrogen is not shown on this graph and will be discussed separately.

by the first addition of a surface active component, much larger concentrations of the same component are required to produce further changes in contact angle. Most of the studies in the literature were conducted on alloys containing 30 to 50 wt% or more of the second component; some studies included the entire range from pure iron to the other pure element. Typically, a composition change of several weight percent is required to produce a measurable change in contact angle. The present study was conducted on samples reasonably close to commercial steel compositions – a few hundredths or tenths of a percent of each alloying component. The addition of a few tenths of a percent of aluminum and/or titanium caused all of the steel samples to have significantly different contact angles than pure iron, but the differences in composition between the various steels were not sufficient to cause significant changes in contact angle from one steel to the next.

Also, the furnace atmosphere did not significantly affect equilibrium contact angle. Steels LC and ULC-Al were each run in pure hydrogen and in forming gas (90 vol% N_2 + 10 vol% H_2). All four runs had equilibrium contact angles within 4° of each other. The only run which did not result in an equilibrium contact angle between 100° – 105° was ULC-AlTi in hydrogen.

Further Investigation of ULC-AlTi

The one steel composition whose contact angle did not lie within the same 5° range as the others was ULC-AlTi (0.002%C, 0.12%Al, 0.12%Ti). When this steel was tested in forming gas, its equilibrium contact angle was lower than the other compositions', but still not significantly different (Figure 3). However, in hydrogen its equilibrium contact angle was significantly lower than any of the other compositions investigated, approximately 90°. To understand this change in contact angle, ULC-AlTi was investigated in atmospheres containing various hydrogen-to-nitrogen ratios. The results of this series of experiments are shown in Figure 4. In an effort to produce a contact angle between the hydrogen value of 90° and the forming gas value of 100°, the experiment was repeated using a mixture of hydrogen and forming gas. The hydrogen flow was set at 100 cm^3/min and the forming gas at 25 cm^3/min using mass flow controllers. With this 82vol% hydrogen atmosphere, the equilibrium contact angle was in the same 5° range as all of the other steels in forming gas. To see if the contact angle could be forced below 90°, the 99.99% pure hydrogen that was normally used in the furnace was passed through a Millipore filter, producing 99.99999% hydrogen. This filter removes moisture, oxygen, CO, and CO_2 to 1 ppb. The equilibrium angle was 90°, virtually unchanged from that produced in standard 99.99% hydrogen, implying that the tiny amount of oxygen and moisture present in the standard grade does not significantly affect the experiment.

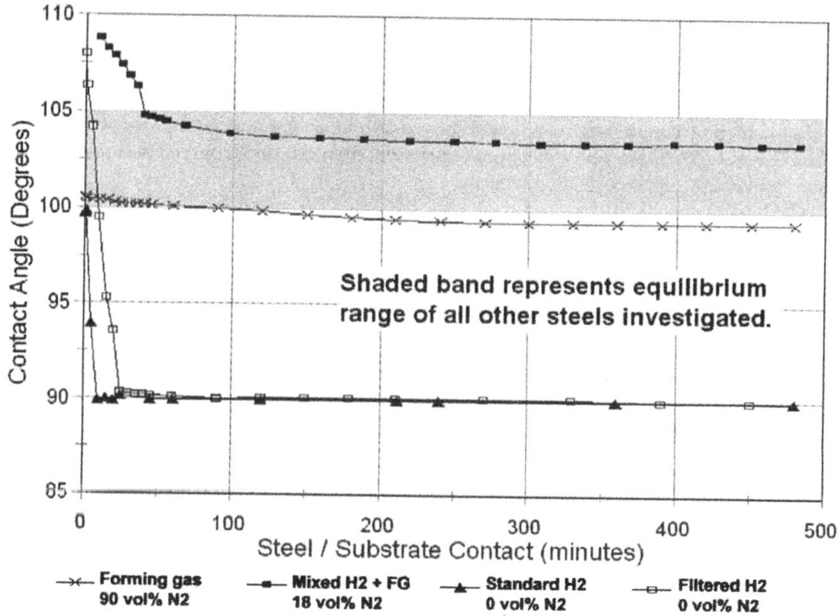

Figure 4 – Contact angles of ULC-AlTi samples on sapphire in atmospheres with various percentages of nitrogen.

The Effect of Atmospheric Nitrogen on Contact Angle

In order to better understand the atmosphere-dependent wetting of steels containing aluminum and titanium, a series of experiments was performed in which the contact angle of a drop was measured as the furnace atmosphere was changed. A drop of steel containing aluminum and/or titanium was formed in an atmosphere of pure hydrogen, then nitrogen was introduced to the atmosphere in a stair-step fashion, as described previously. Four types of steel were used in these experiments: ULC-AlTi and MC-AlTi, which contain similar amounts of aluminum and titanium but have different carbon levels; ULC-AlAl, which contains as much aluminum as the total Al+Ti content of ULC-AlTi; and ULC-Ti, which has the same Ti content as ULC-AlTi. (It was not practical to create a double-Ti steel, "ULC-TiTi", because such high titanium levels are never used in industry.) Exact compositions are listed in Table I. To verify that the atmosphere dependence seen in Figure 4 is actually due to nitrogen, one experiment was run with ULC-AlTi in which argon was introduced into the predominantly hydrogen atmosphere instead of nitrogen, using the stair-step pattern shown in Figure 2. All other experimental conditions remained the same; the substitution of argon for nitrogen was the only difference.

Figure 5 shows the results of this series of experiments. Both steels containing only a single additive did not change contact angles as a function of atmospheric nitrogen. ULC-Ti slowly decreased from 94° to 92° over the course of the experiment, and ULC-AlAl stayed constant at about 104°. Also, ULC-AlTi did not change contact angle as argon was added, but remained constant at about 88°. In the presence of nitrogen, however, the steels that contained both aluminum and titanium monotonically increased contact angle as a function of atmospheric nitrogen content. This interaction between aluminum, titanium, and nitrogen in liquid steel has not been previously reported in the literature.

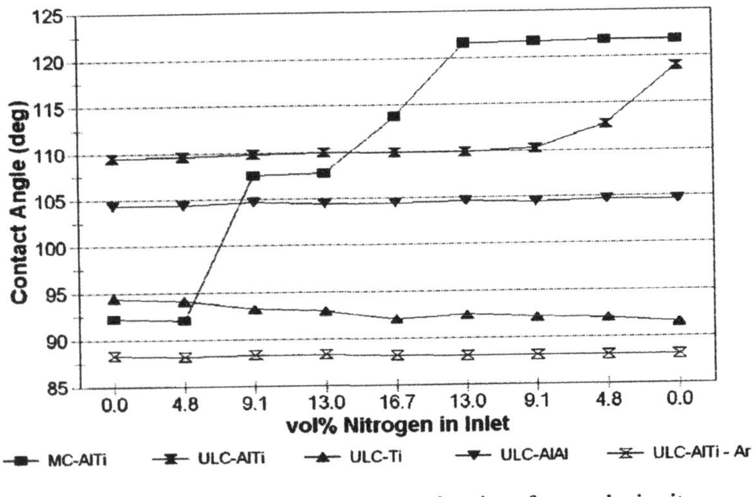

Figure 5 – Steel / Sapphire contact angle as a function of atmospheric nitrogen.

Thermodynamic Considerations of Steel ULC-AlTi in Nitrogen-Containing Atmospheres

Titanium is known to be more surface active in liquid iron than aluminum. Keene[3] reports activity coefficients of

$$\gamma_{Fe\text{-}Al} = \gamma_{Fe} - 18\ mNm^{-1}\cdot(at.\%\ Al) \quad \ldots[2]$$

$$\gamma_{Fe\text{-}Ti} = \gamma_{Fe} - 230\ mNm^{-1}\cdot(at.\%\ Ti) \quad \ldots[3]$$

These values explain the relative positions of the ULC-AlAl and ULC-Ti curves in Figure 5. The titanium addition decreases the liquid iron's surface tension significantly more than the (larger) aluminum addition, resulting in a lower equilibrium contact angle. However, titanium is a very powerful nitride former. When nitrogen is present in the atmosphere, the stability of solid titanium nitride must be considered, as well as how the interactions of the various elements in the steel affect that stability. The calculations throughout this section are based on the work of Sharan, Jimbo, and Cramb[4] and Ozturk and Fruehan[5].

First, consider the case where the steel contains titanium only. The reaction between titanium and nitrogen in liquid iron to form solid titanium nitride is represented by the equilibrium

$$Ti\ (1wt\%) + N\ (1wt\%) \leftrightarrow TiN\ (s) \quad \ldots[4]$$

where standard states for titanium and nitrogen are 1 wt%. The equilibrium constant for this reaction, K, is

$$K = \frac{a_{TiN}}{a_{Ti}a_N} = \frac{1}{\%Ti\ f_{Ti}\ \%N\ f_N} \quad \ldots[5]$$

where a_i and f_i are the activity and activity coefficient of the ith component, respectively. The activity coefficients are calculated from first-order interaction coefficients, e_i^j, available in the literature, using equations of the form:

$$\log f_i = e_i^j[\%j] \quad \ldots[6]$$

Table II lists the first-order interaction coefficients relevant to the present work. These data are from Sigworth and Elliott[6], with the exceptions noted[7,8,9]. Using the Sigworth and Elliott data, the activity coefficients for Ti and N are

$$\log f_{Ti} = 0.013\ (\%Ti) - 1.8\ (\%N) \quad \ldots[7]$$

$$\log f_N = -0.053\ (\%Ti) \quad \ldots[8]$$

Sharan, Jimbo, and Cramb[4] report the TiN equilibrium constant, K, to vary with temperature (in °K) according to the equation

$$\ln K = \frac{37140}{T} - 13.75 \quad \ldots[9]$$

Combining equations 5 through 9, the equilibrium nitrogen concentration, %N, for a given titanium concentration, %Ti, at temperature T can be calculated from the equation

$$\log(\%N) + e_{Ti}^{N}(\%N) = -\log(\%Ti) - (\%Ti)(e_{Ti}^{Ti} + e_{N}^{Ti}) - \frac{1}{\ln 10}\left(\frac{37140}{T} - 13.75\right) \quad \ldots[10]$$

Equation 10 and the temperature-dependence equations in Table IIB were used to calculate the stability diagram for TiN at 1550, 1575, and 1600°C, shown in Figure 6. At the experimental temperature of 1600°C, Figure 6 shows that steel ULC-Ti (0.1% Ti) has an equilibrium nitrogen content of 0.03%. Any higher concentration of nitrogen in the steel will cause precipitation of solid TiN.

Tables II A and B – First-order interaction coefficients for various elements in liquid iron, and their temperature dependence.

i↓ j→	Al	C	N	Ti
Al	0.045	0.091	-0.058	0.0040*
C	0.043	0.14	0.11	-0.049**
N	-0.028	0.13	0.0	-0.53
Ti	0.0037***	-0.21**	-1.8	0.013

A) First-order interaction coefficients for Al, C, N, and Ti in liquid iron at 1873K, from Sigworth and Elliott[6] except as noted.
General form of equation for activity coefficient f_i is: $\log(f_i) = e_i^j[\%j]$
*from Li and Suito[7]
**from *Steelmaking Data Sourcebook*[8]
***from Mutale, Pistorius, and Geldenhuis[9]

i↓ j→	Al	C	N	Ti
Al	63/T + 0.011	–	1650/T - 0.94	–
C	–	158/T + 0.0581	–	–
N	859/T -0.487	–	–	-4070/T + 1.643
Ti	–	–	-13900/T + 5.61	–

B) Temperature dependence of first-order interaction coefficients e_i^j (T in °K)

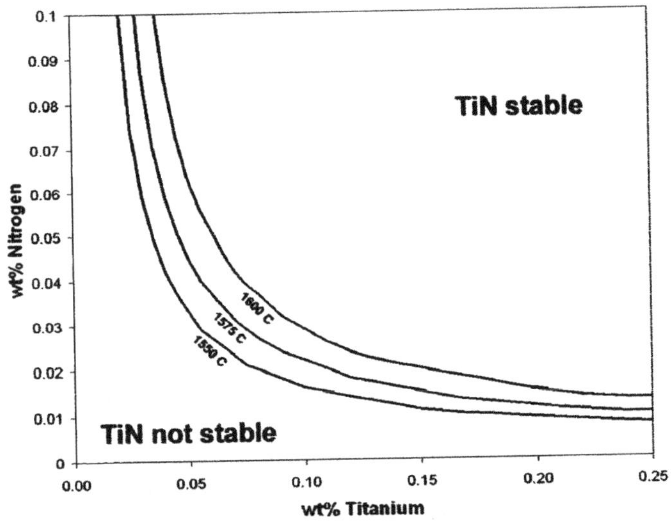

Figure 6 – Stability of titanium nitride as a function of temperature.

The data from Table IIA also allow the effect of carbon content to be calculated. Considering Ti and C, the terms f_{Ti} and f_N become

$$\log(f_{Ti}) = \%C\, e_{Ti}^{C}\ \%N\, e_{Ti}^{N}\ \%Ti\, e_{Ti}^{Ti} \qquad ...[11]$$

$$\log(f_N) = \%C\, e_{N}^{C}\ \%Ti\, e_{N}^{Ti} \qquad ...[12]$$

Using equations 11 and 12 and the temperature dependence of K used earlier, the equilibrium concentration of nitrogen can be written

$$\log(\%N) + e_{Ti}^{N}(\%N) = -\log(\%Ti) - (\%Ti)(e_{N}^{Ti} + e_{Ti}^{Ti}) - (\%C)(e_{N}^{C} + e_{Ti}^{C}) - \frac{1}{\ln 10}\left(\frac{37140}{T} - 13.75\right) \qquad ...[13]$$

Substituting the numerical values of the interaction coefficients, at 1873K

$$\log(\%N) - 1.8(\%N) = -\log(\%Ti) + 0.517(\%Ti) + 0.08(\%C) - 2.64 \qquad ...[14]$$

Equation 14 was used to calculate the Ti-N stability curve at 1873K for several carbon contents, shown in Figure 7.

Figure 7 shows that as the carbon content of the steel increases, the equilibrium nitrogen content slightly increases. Therefore, at any given point in the course of the experiment, steel MC-AlTi (0.3% C) has slightly more dissolved nitrogen than steel ULC-AlTi (0.003% C). From Table IIA, we see that nitrogen decreases the activity of both aluminum and titanium in steel ($e_{Al}^{N} = -0.058$, $e_{Ti}^{N} = -1.8$), but interacts more with titanium than aluminum (larger negative interaction coefficient). This means that as nitrogen is added to the system, the activity of titanium will be reduced more than the activity of aluminum. In Figure 5, as nitrogen is added to the atmosphere the contact angle of MC-AlTi shifts from that of ULC-Ti to that of ULC-AlAl.

Furthermore, carbon itself reduces the activity of titanium ($e_{Ti}^{C} = -0.21$) but *increases* the activity of aluminum ($e_{Al}^{C} = 0.091$) and of nitrogen ($e_{N}^{C} = 0.13$). Therefore, in both systems the activity of titanium decreases as nitrogen is introduced to the system, but it is decreased more in MC-AlTi than in ULC-AlTi. This explains the monotonic increase in contact angle in both MC-AlTi and ULC-AlTi, and why MC-AlTi showed a larger change in angle than ULC-AlTi.

Despite the reduction in titanium activity due to the presence of carbon and nitrogen, both -AlTi steels were exposed to sufficient nitrogen for an extended period of time to eventually form solid TiN, according to the equilibrium in equations 4 and 5. As solid titanium nitride precipitated, the concentration of Ti dissolved in the steel decreased, causing an increase in γ_{LV} and resulting in a gradual increase in the contact angle. This explains why the contact angle continued to increase even when atmospheric nitrogen was decreased. Titanium nitride is quite stable, and an extended soak in pure hydrogen would be required to purge the system of nitrogen sufficiently to drive the equilibrium of equation 4 back to the left.

Figure 8 shows a precipitate found near the solid/liquid/vapor interface of a ULC-AlTi sample. The pictures on the top row are elemental maps, which are computer generated images showing which elements were detected in each pixel when a scanning EDS analysis was performed across the sample's surface; on the bottom row are the SEM image and a composite of the elemental maps. Note that EDS cannot detect nitrogen reliably, so only metals and oxygen are shown. However, the precipitate must be a nitride because elemental mapping did not show oxygen in the precipitate and titanium carbide or carbonitride will not form above the steel liquidus temperature[4]. Although Figure 8 shows this precipitate to be a mixed cation aluminum/titanium nitride instead of pure TiN, the line of reasoning presented previously is still valid – surface active elements precipitated out of the steel, leading to an increase in contact angle.

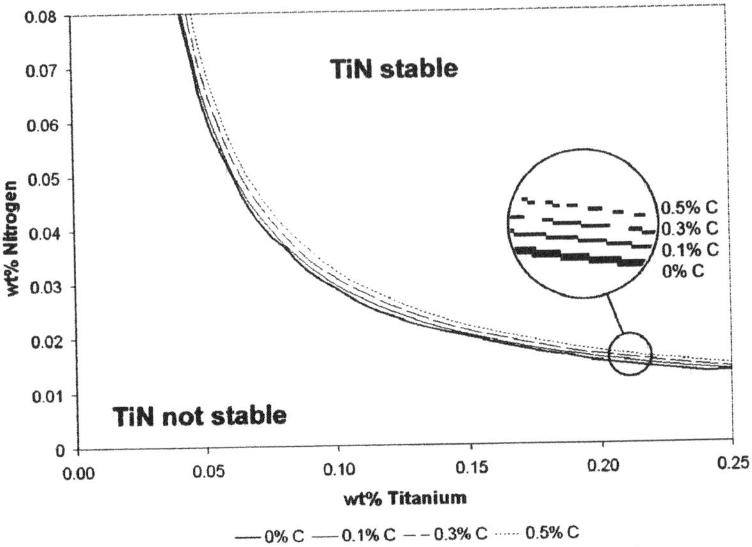

Figure 7 – Stability of titanium nitride at 1873K as a function of carbon content.

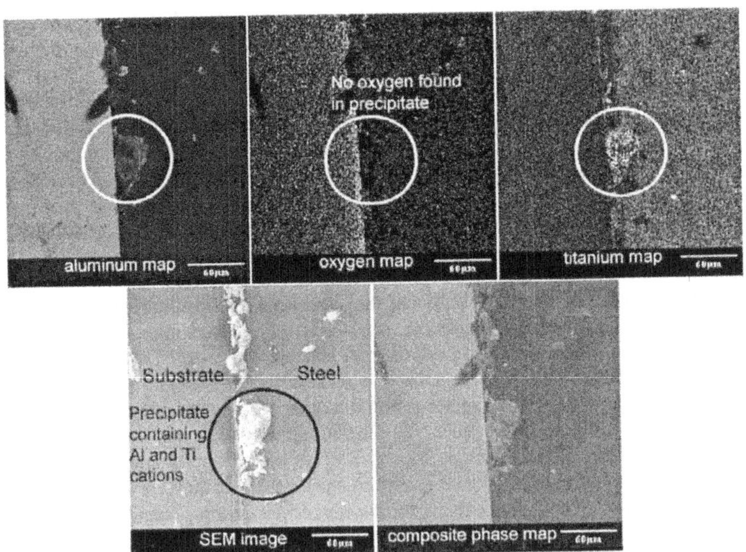

Figure 8– Elemental map of aluminum / titanium nitride precipitate near edge of ULC-AlTi drop.

CONCLUSIONS

1) Each of the steel compositions tested had equilibrium contact angles significantly lower than pure iron, but there were not significant differences between the individual steels, with the exception of ULC-AlTi in pure hydrogen. The low concentrations of aluminum and/or titanium in the steels caused a large decrease in the solid/liquid interfacial energy resulting in lower contact angles than pure iron. However, there were not enough differences in the steel compositions to result in measurable differences in contact angles between them. Further decreases in the contact angle would require the addition of higher concentrations of aluminum or titanium than are used in steelmaking. Lower contact angles were reported in the literature only when aluminum or titanium were added in concentrations at least an order of magnitude higher than those used in the present work. In conclusion, changes in the concentration of aluminum and titanium within normal steelmaking practices have little effect on contact angle. Therefore, manipulation of wetting by changing the aluminum or titanium content can be eliminated as a potential anti-clogging technique.

2) Steels containing both aluminum and titanium showed a monotonic increase in contact angle as a function of atmospheric nitrogen content, particularly at higher carbon levels. This phenomenon was explained in terms of the first-order interaction coefficients of titanium, nitrogen, and carbon in liquid iron. Nitrogen causes a decrease in the activity of titanium, resulting in an increase in contact angle. Increasing carbon content enhances the effect of nitrogen, causing a larger increase in contact angle. When sufficient nitrogen was added to the system, solid aluminum/titanium nitride precipitates formed. As these precipitates formed, the concentration of titanium and aluminum dissolved in the steel decreased, leading to increases in contact angle even when the atmospheric nitrogen content decreased.

REFERENCES

[1] S. Ramachandran. "Static Nozzle Accretion Study to Understand Nozzle Clogging During Continuous Casting," M.S. Thesis, University of Missouri-Rolla (2001).

[2] P. D. Ownby and K.W.K. Li, *J. of the American Ceramic Society*, **74** [6], 1275 (1991).

[3] B. J. Keene. "Review of Data for the Surface Tension of Iron and Its Binary Alloys," *International Materials Reviews*, **33** [1], 1 (1988).

[4] A. Sharan, I. Jimbo, A.W. Cramb. "Fundamental Aspects of the Casting of Titanium Treated Steels," *ISS Transactions*, **16**, 95 (1995).

[5] B. Ozturk and R.J. Fruehan. "Thermodynamics of Inclusion Formation in Fe-Ti-C-N alloys," *Metallurgical Transactions B*, **21B** [October], 879 (1990).

[6] G.K. Sigworth and J.F. Elliott. "The Thermodynamics of Liquid Dilute Iron Alloys," *Metallurgical Science*, **8**, 298 (1974).

[7] G. Li and H. Suito. "Galvanic Cell Measurements on Supersaturated Activities of Oxygen in Fe-Al-M (M = C, Te, Mn, Cr, Si, Ti, Zr and Ce) Melts", *Metallurgical and Materials Transactions B*, **28B**, 251 (1997).

[8] *Steelmaking Data Sourcebook*, The Society for the Promotion of Science, The 19th Committee Steelmaking, Gordon and Breach Publishers, New York, NY, 1988.

[9] C.T. Mutale, P.C. Pistorius and J.M.A. Geldenhuis. "Electrochemical Measurements of Oxygen Activities in Aluminum and Titanium Containing Stainless Steel at 1600 °C", submitted to Ironmaking and Steelmaking.

QUANTITATIVE PHASE ANALYSIS OF REACTIVE WETTING OF MULLITE BY SILICATE GLASSES

P. Darrell Ownby and Benjamin T. Eldred
University of Missouri – Rolla
Department of Materials Science and Engineering
222 McNutt Hall
Rolla, MO 65401

Wolfgang Braue and Bernd Hildmann
German Aerospace Center (DLR)
Materials Research Institute
D-51147 Cologne, Germany

ABSTRACT

A new method of using EDS analysis to quantify reactive wetting has been developed. By setting up equations based on conservation of mass and EDS chemical analysis and using Mathcad or similar software to solve the system of simultaneous equations, molar coefficients of reaction between the liquid drop and the substrate can be calculated. Experiments were conducted with a non-reactive borosilicate glass and a highly reactive yttrium-aluminosilicate glass on single crystal and polycrystalline mullite. In both systems, the calculated molar coefficients of reaction agreed well with previously published information about the surface energy of various mullite planes and with traditional sessile drop wetting data.

INTRODUCTION

Recently it has been shown that the toughness of mullite can be significantly improved by using an intergranular glassy layer to induce elongated grain growth[1]. Huang et al[2] published a study of the wetting behavior of two glasses, one borosilicate and one yttrium-aluminosilicate (YAS), potentially suited for use as toughening agents in polycrystalline mullite. It was found that both glasses wet polycrystalline mullite, but the YAS glass wet significantly better than the borosilicate and showed more penetration of the grain boundaries. Also, the YAS glass reacted with the substrate to produce a complex morphology with several different products, but the borosilicate glass showed little reaction.

This paper presents a study of the same glasses wetting single crystal mullite. The absence of grain boundaries significantly impacts the interaction between both glasses and the mullite, particularly the highly reactive YAS glass. Balanced equations generated from EDS data show how various crystal planes interact differently with the glasses. The equilibrium contact angle also shows a strong dependence on the crystal plane, in agreement with previous studies of mullite surface energy.

SEM / EDS ANALYSIS OF BOROSILICATE GLASS

SEM Analysis of Borosilicate Glass.

In all experiments, a piece of glass weighing about 0.02 g was melted on a polished section of mullite using standard sessile drop techniques documented elsewhere[3]. A digital video camera was used to film the drop throughout the course of each experiment, and the contact angle was measured from the digital images. Afterwards the drops were sectioned for SEM and analyzed using EDS parameters appropriate for detection of light elements such as boron. The single crystal substrates were provided by the Institute of Materials Research at the German Aerospace Center.

Figure 1 shows micrographs of borosilicate glass on each of four mullite substrates – (100), (010), (001), and polycrystalline. A - L are the locations of EDS analysis. In all cases, varying amounts of mullite dissolved in the molten glass then recrystallized as elongated crystals on cooling. On all single crystal substrates, the recrystallized mullite preferentially aligned perpendicular to the (001) plane of the substrate mullite. Figure 1C shows this alignment directly; in 1A, the plane of the micrograph includes the (001) axis, so the recrystallized mullite is shown in cross-section. The glass / polycrystalline interface showed significantly more dissolution of the substrate by the glass than the glass / single crystal interfaces due to the presence of grain boundaries. Despite the relatively poor wetting of borosilicate glass on mullite, which will be addressed in a later section, grain boundaries enhance the corrosion rate, so more mullite was dissolved and more crystals were found precipitated in the solidified drop.

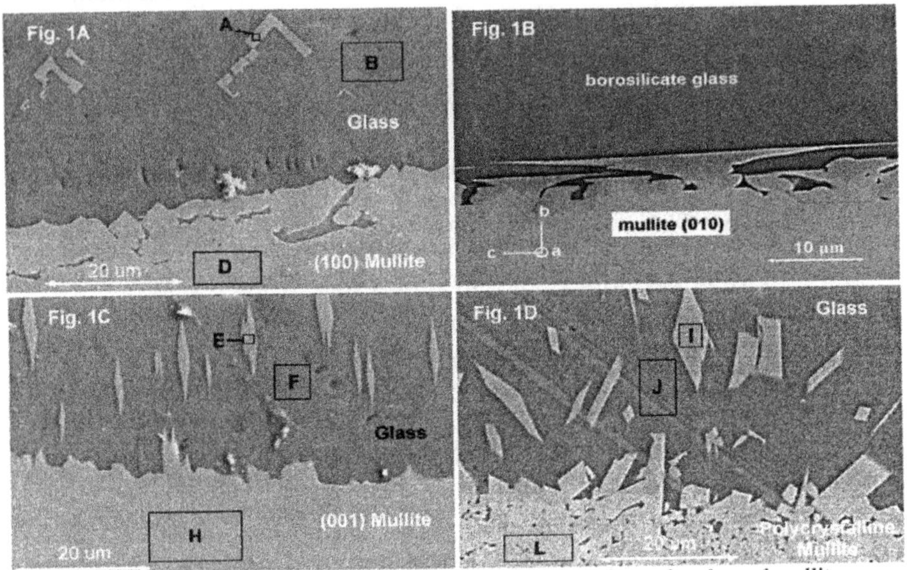

Figure 1 – Borosilicate glass on mullite. All single crystal samples showed mullite recrystallization, preferentially perpendicular to the (001) substrate surface.

Reaction Model for EDS Analysis.

EDS analysis was performed to identify the various phases visible throughout the cross-section of each drop. It was assumed that the glass near the interface dissolved mullite from the substrate, altering the chemistry of the glass (not of the substrate), and then precipitated alumina-rich mullite crystals upon cooling. It was also assumed that excess mullite was available for reaction. The third assumption of the model was that the dissolved mullite did not diffuse all the way to the top of the drop, leaving an unreacted region of glass which retained the original bulk composition far from the interface. The boundary between the "near" and "far" regions could be roughly approximated by observing the extent of precipitated crystals in the cross-section. However, the exact determination of reacted and unreacted glass was left to computer analysis of the EDS data, as discussed in the following sections. Four distinct areas, illustrated in Figure 2, were tested in each sample: the precipitated crystals on the glass side of the interface; the reacted glass around the crystals, near the interface; the unreacted glass at the top of the drop, far from the interface; and the substrate.

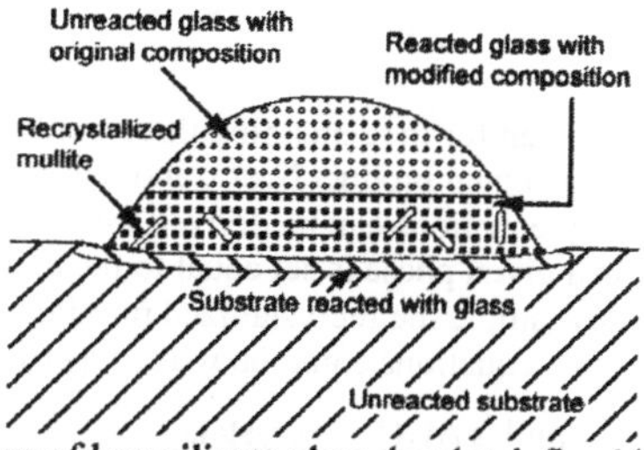

Figure 2 – Regions of borosilicate glass droplet defined in reaction model

EDS Analysis for Borosilicate Glass on Mullite.

The positions which were analyzed are labeled A-L in Figure 1, and the compositions are listed in Table I. A small amount of surface carbon was detected on each sample, but the data in Table I has been normalized to 100% metal oxides, eliminating this known contamination.

Table I – EDS analysis of borosilicate glass / mullite interface.

Location	Description	mol% Al_2O_3	mol% SiO_2	mol% B_2O_3	mol% Na_2O
A	(100) L-shaped crystal in glass	73.8	26.2		
B	(100) Glass near interface (near "A")	8.0	70.0	16.4	5.6
C*	(100) Glass near top of drop	2.0	77.6	16.4	4.1
D	(100) Substrate	68.7	31.3		
E	(001) Diamond-shaped crystal in glass	72.0	28.0		
F	(001) Glass near interface (near "E")	8.1	70.4	16.5	5.0
G*	(001) Glass near top of drop	7.8	75.0	13.6	3.6
H	(001) Substrate	69.0	31.0		
I	Polycrystalline. Diamond-shaped crystal in glass	72.4	27.6		
J	Polycrystalline. Glass near interface (near "I")	4.8	74.7	16.5	4.0
K*	Polycrystalline. Glass near top of drop	7.1	75.6	13.9	3.4
L	Polycrystalline. Substrate	67.6	32.4		
glass	Starting glass composition	1.3	77.4	16.8	4.5

*Locations C, G, and K are not shown on the micrographs. They are located at the top of the drops, far from the reaction zone.

For each sample, a variable was assigned to each oxide reactant and product – moles of SiO_2 in the starting glass, moles of B_2O_3 in the starting glass, moles of SiO_2 in the recrystallized mullite, and so forth. Four equations can be written from conservation of mass. For silica,

SiO_2 in starting glass + SiO_2 from substrate = SiO_2 in glass near interface + SiO_2 in glass far from interface + SiO_2 in recrystallized mullite ...[1]

And similar equations can be written for the remaining three oxides: reactant B_2O_3 must equal product B_2O_3, reactant Na_2O must equal product Na_2O, and reactant Al_2O_3 must equal product Al_2O_3. The remaining necessary equations can be derived from EDS data. In each case, the molar fraction of each oxide in a given phase is set equal to the amount of that oxide detected by EDS. For example, in the (100) sample the recrystallized mullite was found to be composed of 74% Al_2O_3 and 26% SiO_2, so two equations could be written as follows:

$$\frac{\text{moles } Al_2O_3 \text{ in crystals}}{\text{moles } Al_2O_3 \text{ in crystals} + \text{ moles } SiO_2 \text{ in crystals}} = 0.74 \qquad ...[2]$$

$$\frac{\text{moles } SiO_2 \text{ in crystals}}{\text{moles } Al_2O_3 \text{ in crystals} + \text{ moles } SiO_2 \text{ in crystals}} = 0.26 \qquad ...[3]$$

Similarly, the glass near the interface of (100) contained 16.4% B_2O_3, so

$$\frac{\text{moles } B_2O_3 \text{ in glass near interface}}{\text{moles } B_2O_3 \text{ in glass} + \text{moles } SiO_2 \text{ in glass} + \text{moles } Al_2O_3 \text{ in glass} + \text{moles } Na_2O \text{ in glass}} = 0.164 \qquad ...[4]$$

Similar equations written for the other oxide reactants and products produced a system of twenty equations with sixteen variables. Mathcad 2000 Pro software was used to solve this system of 20 equations with sixteen variables. The initial "guess" values required by the solving engine were based on observations of the micrographs – amount of precipitated crystals, how much of the drop appeared to be reacted, etc. For each system of equations, the initial values were varied slightly to ensure that the solving engine consistently converged to the same solution. Mathcad calculated the value of each oxide variable, producing an equation of the form

reactant oxide + reactant oxide + ... → product oxide + product oxide + ...

Since the composition of each reactant and product is known, the calculated number of moles of each oxide can be factored to produce an equation of the form

original glass + mullite → glass near interface + glass far from interface + recrystallized mullite

For example, for borosilicate glass reacting with (100) mullite, Mathcad calculated the following balanced chemical reaction:

674.0(as-received glass) + 47.4(mullite substrate) →
120.9(glass near interface) + 573.1(glass far from interface) + 27.3(recrys. mullite) ...[5]

674.0 (0.774 SiO_2 + 0.168 B_2O_3 + 0.045 Na_2O + 0.013 Al_2O_3) +
47.4 (0.687 Al_2O_3 + 0.313 SiO_2) →
120.9 (0.700 SiO_2 + 0.164 B_2O_3 + 0.056 Na_2O + 0.080 Al_2O_3) +
573.1 (0.776 SiO_2 + 0.163 B_2O_3 + 0.041 Na_2O + 0.020 Al_2O_3) +
27.3 (0.738 Al_2O_3 +0.262 SiO_2) ...[6]

Due to rounding and approximations in the solving engine, the calculated chemical reaction did not always balance exactly. In these cases, the number of moles of each product was adjusted slightly until a balanced equation was obtained. Finally, the reaction was normalized to 100 moles of starting glass by multiplying both sides of the equation by a constant.

The error inherent to Mathcad's solving engine can be considered negligible compared to the error in the EDS data. For most elements, the reported percentage of each element is ±5%; for light elements like boron, this can increase to ±10%. However, since the EDS reports each oxide as a weight percentage of the total oxides present, normalized to 100%, the error in the reported percentages is actually a function of the uncertainty in all the oxides present. Assuming 10% error in B_2O_3 and 5% in all other oxides, the uncertainty in reported percentages calculates to approximately 6%. Therefore, reported percentages and calculated reaction coefficients in this paper can be assumed to be about ± 6%.

Using the same method demonstrated for the (100) sample, balanced equations were written for the (010), the (001), and the polycrystalline samples. The coefficients for each of these equations were normalized to 100 moles of starting glass by multiplying both sides of each equation by a constant. The normalized coefficients are presented in Table II.

Table II – Normalized molar coefficients of reaction between borosilicate glass and mullite.

	Starting Glass	Substrate	Glass Near Interface	Glass Far From Interface	Recrystallized Mullite
(100), normalized to 100 moles	100	7.0	18.0	85.0	3.0
(010), normalized to 100 moles	100	6.8	23.3	82.1	3.2
(001), normalized to 100 moles	100	13.5	59.5	51.0	2.5
polycrystalline, normalized to 100 moles	100	32.8	56.9	51.8	23.5

The (100) sample dissolved the least amount of mullite and showed the least diffusion of alumina into the bulk glass at the top of the drop (see Table I). In this sample, the alumina stayed segregated near the interface, producing a small alumina-rich zone with recrystallized mullite. The (001) sample dissolved twice as much mullite as the (100) (see Table II) and the alumina dispersed more uniformly throughout the drop (see Tables I & II). This can be inferred from the similar amounts of "near" and "far" glass (Table II), and from the fact that less mullite recrystallized even though more was dissolved. The polycrystalline sample behaved similarly to the (001), showing similar amounts of "near" and "far" glass, except that more than twice as much mullite was dissolved, and almost ten times as much recrystallized. From Table I, the polycrystalline sample was the only one to show more alumina in the bulk glass far from the interface than in the glass near the interface. Apparently, this sample first dissolved a large excess of mullite and achieved a fairly uniform enrichment of alumina throughout the drop, very similar in composition to the (001) sample. Then the excess alumina was precipitated in the recrystallized mullite near the interface, actually depleting the interfacial region in alumina relative to the newly established equilibrium in the glass drop.

SEM / EDS ANALYSIS OF YAS-6 GLASS

In contrast to the borosilicate glass, the yttria-aluminosilicate glass (YAS-6) reacted strongly with both single crystal and polycrystalline mullite substrates, and also devitrified on cooling. Figure 3 shows low magnification images of cross-sections of the (100) and polycrystalline samples. The entire volume of the drops on the (100), (001), and polycrystalline substrates crystallized on cooling to form large mullite and corundum crystals suspended in a matrix of $Y_2O_3{\bullet}2SiO_2$ and a SiO_2-rich aluminosilicate with traces of residual glass. Similar morphology was found in the (010) sample, but only part of the drop crystallized – some glass still remained at the top of the drop. In the polycrystalline sample, nodules of $Y_2O_3{\bullet}2SiO_2$ were also found between the grains of the substrate near the solid/liquid interface, and glass had visibly penetrated the substrate to a depth of nearly a millimeter.

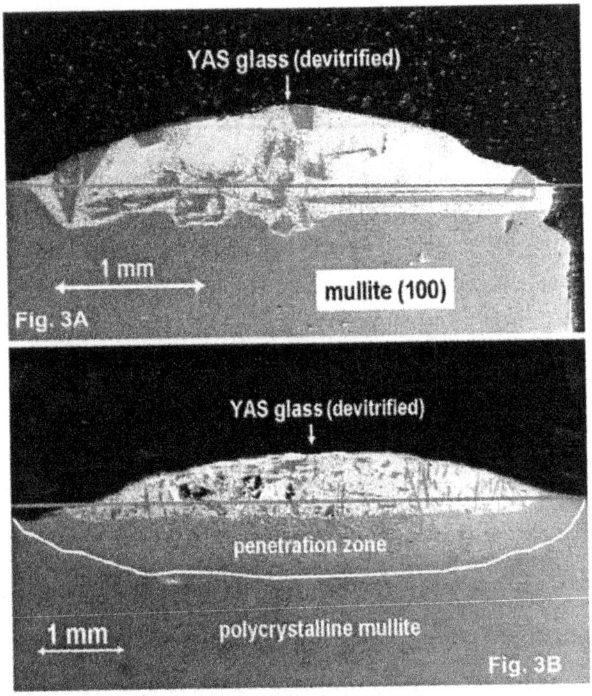

Figure 3 – Devitrified YAS glass showing strong reaction with mullite substrates.

The extensive penetration of glass into the polycrystalline substrate is directly attributable to the presence of grain boundaries. As reported in a previous paper[2] and verified in this work, YAS glass wets mullite very well, with contact angles down to 6° at 1600°C. Such a low contact angle makes it very likely that the glass will penetrate the grain boundaries of the polycrystalline sample. Thus, during the experiment each mullite grain in the reaction zone is attacked on many sides by an intergranular film of glass. This leads to a high dissolution rate without moving the visible interface line.

Figure 4 shows micrographs of the various phases found in the devitrified assemblages. In all samples, large corundum plates were found to have grown throughout the drop with fine $Y_2O_3 \cdot 2SiO_2$ crystals between them. The solid/liquid interface of some samples showed strong serration aligned with the growth planes of the corundum. EDS was performed at locations M-W indicated on the micrographs and the results are presented in Table III, normalized to 100% to account for known surface carbon contamination. The $Y_2O_3 \cdot 2SiO_2$ detected in EDS has been confirmed by its zone axis electron diffraction pattern in TEM. Electron diffraction data also indicates that the darker phase (Q) is crystalline as well, although its precise identification is still in question. Detailed TEM investigation of this phase is currently underway.

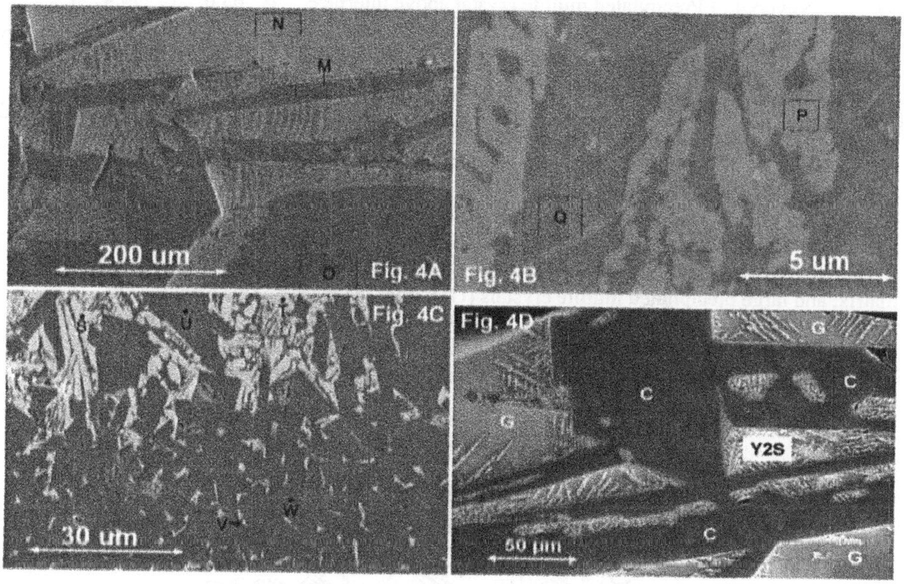

Figure 4 – EDS analysis of various phases in devitrified YAS glass.
4A – Interfacial region of (010) sample.
4B – Close up of $Y_2O_3 \cdot 2SiO_2$.
4C – Interfacial region of polycrystalline sample.
4D – Corundum (C), residual glass (G), and $Y_2O_3 \cdot 2SiO_2$ in (100).

Table III – EDS analysis of YAS-6 glass / mullite interface.

Location	Substrate	Description	mol% Al_2O_3	mol% SiO_2	mol% Y_2O_3
M	(010)	Long alumina crystal on glass side of interface	100.0		
N	(010)	Gray glass above M	27.1	56.6	16.3
O	(010)	Substrate	67.8	32.2	
P	(010)	Close up of striations between alumina crystals; crystalline α-Y_2O_3•$2SiO_2$	4.7	60.9	34.4
Q	(010)	Close up of striations between large crystals; crystalline, exact phase undetermined	31.8	60.4	7.8
R*	(010)	Glass at top of drop	26.5	57.5	16.0
S	polycrys.	Crystalline Y_2O_3•$2SiO_2$ in matrix above interface	3.7	64.1	32.2
T	polycrys.	Crystalline Y_2O_3•$2SiO_2$ in matrix above interface	1.9	64.3	33.8
U	polycrys.	Precipitated mullite crystal above interface	65.6	34.4	
V	polycrys.	Crystalline Y_2O_3•$2SiO_2$ precipitated between mullite grains below the interface	11.2	61.6	27.2
W	polycrys.	Mullite substrate	64.3	35.7	
glass	---	Starting glass composition	25.0	56.0	19.0

*Location R, located at the top of the drop in the unreacted glass, is not shown on the photos.

Reaction Model for EDS Analysis.

For EDS analysis, it was assumed that the YAS-6 glass reacted with mullite from the substrate, then upon cooling produced a crystalline matrix consisting of Y_2O_3•$2SiO_2$ (S,T Table III) and a SiO_2-rich aluminosilicate (Q), large precipitated alumina crystals (M) within the matrix, and large precipitated mullite crystals (U) within the matrix. These phases are represented schematically in Figure 5. Unreacted glass was only included in the case of the (010) substrate, as it was only found in trace quantities in other samples. In the case of the polycrystalline substrate, the small nodules of Y_2O_3•$2SiO_2$ visible in Figure 4C (V, Table III) were also included.

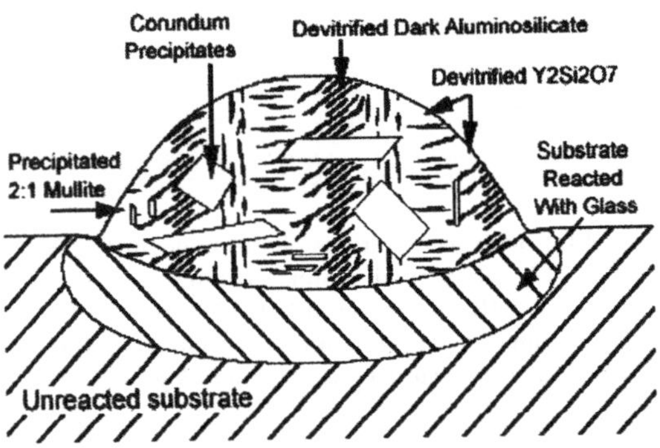

Figure 5 – Reaction model for YAS-6 glass on mullite substrates.

EDS Analysis of YAS-6 Glass on Mullite.

As with the borosilicate glass, a system of equations was set up to balance the reaction between the YAS-6 glass and each of the mullite substrates. Since the crystalline reaction products were positively identified by electron diffraction, the equations were set up in Mathcad using the theoretical molar ratios of each phase. EDS data was used for glassy or uncertain phases. A summary of the phase compositions used to calculate the reaction between YAS-6 glass and each of the mullite substrates is given in Table IV.

Table IV – Compositions of reactants and products used to calculate the interaction between YAS-6 glass and mullite substrates.

Description	Reactant	Product	mol% Al_2O_3	mol% SiO_2	mol% Y_2O_3
Starting glass composition	X		25.0	56.0	19.0
Substrate	X		66.7	33.3	0
Matrix $Y_2O_3{\cdot}2SiO_2$		X	0	66.7	33.3
Dark matrix phase		X	31.8	60.4	7.8
Alumina crystals precipitated in matrix		X	100.0	0	0
Mullite crystals precipitated in matrix		X	66.7	33.3	0
Unreacted glass at top of sample on (010) substrate		X	27.1	56.6	16.3
$Y_2O_3{\cdot}2SiO_2$ precipitated between mullite grains in the polycrystalline substrate		X	0	66.6	33.3

Mathcad calculated a balanced chemical reaction between YAS-6 glass and each of the substrates, using identical methodology as with the borosilicate samples. A summary of the reaction coefficients for YAS glass reacting with mullite is presented in Table V. All of the coefficients have been normalized to 100 moles of starting glass by multiplying both sides of the equation by a constant. As an example, for YAS-6 on (100) mullite,

$$100(\text{starting glass}) + 72.4(\text{substrate}) \rightarrow 53.3(Y_2O_3{\cdot}2SiO_2) + 42.9(\text{recrys. mullite}) + 31.4(\text{corundum}) + 44.8(\text{dark aluminosilicate}) \qquad \ldots[7]$$

Table V – Normalized molar coefficients of reaction between YAS-6 glass and mullite.

	Reactants		Products					
	Starting Glass	Substrate	$Y_2O_3{\cdot}2SiO_2$	Recrys. Mullite	Corundum	dark alumino-silicate	$Y_2O_3{\cdot}2SiO_2$ nodules in substrate	unreacted glass
(100)	100	72.4	53.3	42.9	31.4	44.8	--	--
(010)	100	16.2	22.5	3.6	11.8	16.6	--	62.6
(001)	100	135.0	30.5	39.0	58.0	107.0	--	--
polycrystalline	100	37.1	43.0	16.3	21.2	55.6	1.0	--

CONTACT ANGLE MEASUREMENTS

The contact angle of the borosilicate glass as a function of temperature is shown in Figure 6. All three of the samples on single crystal substrates underwent a gradual softening between about 1300°C and 1450°C, achieving a steady state above 1450°C. The (001) sample softened slightly earlier than the other two and also reached equilibrium earlier, but the shape of the curve is similar to the others.

The contact angle is determined by a balance of forces expressed in the Young Equation:

$$\cos(\theta) = (\gamma_{SV} - \gamma_{SL}) \div \gamma_{LV} \qquad \text{...[8]}$$

where θ is the equilibrium contact angle, and γ_{SV}, γ_{SL}, and γ_{LV} are the solid-vapor, solid-liquid, and liquid-vapor surface energies, respectively. For a wetting system (θ<90), increasing γ_{SV} while keeping other factors constant results in a decrease in contact angle. Previous works[1, 4] have established the (001) plane as the highest energy plane in the mullite system, and the (001) sample did have the lowest equilibrium contact angle, 28°, of the borosilicate samples. (100) had an equilibrium angle of 45°, and (010) had the highest equilibrium angle of 64°. The polycrystalline sample followed a simple rule of mixtures, with an equilibrium angle of 43°. Looking at the molar coefficients of reaction in Table II, we see that the (001) sample, the highest energy plane, is also the most reactive of the single crystal samples.

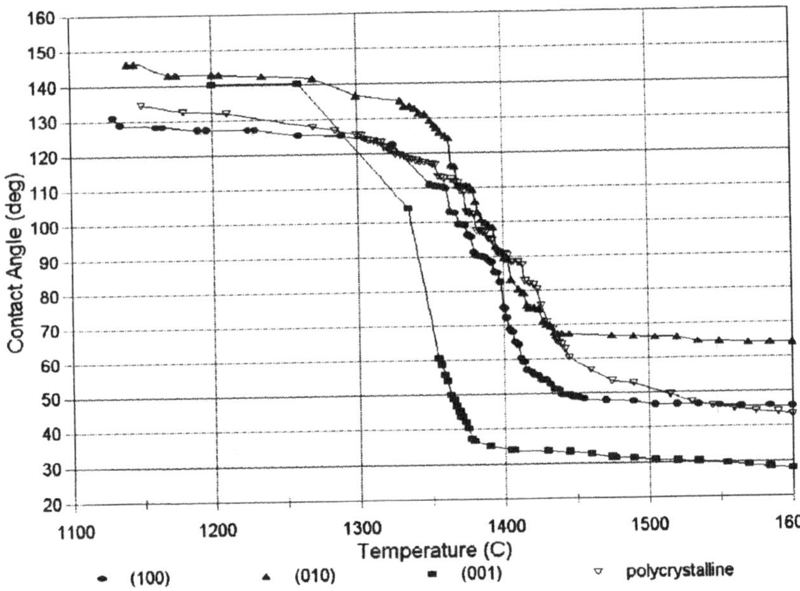

Figure 6 – Contact angle of borosilicate glass on mullite.

Figure 7 shows the contact angle of the YAS glass as a function of temperature. These samples exhibited an almost crystal-like melting, transforming from the angular, as-received pieces to low-contact-angle, molten drops over the temperature range 1400°C-1425°C. Above 1425°C, the (001) sample quickly reached an equilibrium angle of 6°, but the other samples continued to decrease in contact angle throughout the experiment. The (100) and (010) samples reached contact angles of 19° and 28°, respectively, at 1600°C, and the polycrystalline sample again showed rule of mixtures behavior with a final contact angle of 16°.

The wetting phenomena of YAS glass on mullite cannot be adequately explained by classic wetting theory such as the Young-Dupré equation. Since the composition of the liquid in the interfacial region is constantly changing due to dissolution of the substrate, local chemical reactions, and precipitation of refractory products, the local surface energies can be very different from the surface energies of the bulk, as-received glass. Reactive wetting of ceramics by metals has been extensively investigated[5, 6, 7, 8, 9] and it is generally concluded that extra terms must be added to the Young-Dupré equation to account for the additional energy of reaction. Most commonly, the reaction will decrease the solid-liquid interfacial energy, leading to a decrease in the contact angle, however reactions that increase γ_{SL} and the contact angle have also been detected. The same principles apply in this case – the reaction between the glass and the substrate decreased the interfacial tension and caused a significant decrease in the contact angle. However, in both the borosilicate and YAS systems, the contact angles followed the same pattern: the (001) substrate had the lowest angle, followed by the polycrystalline, (100), and (010) consecutively. Even though the wetting behavior of the YAS system is dominated by reactive wetting, the influence of the solid-liquid and solid-vapor surface energies is evident in the significantly different contact angles established on each of the single crystal substrates.

Figure 7 – Contact angle of YAS-6 glass on mullite.

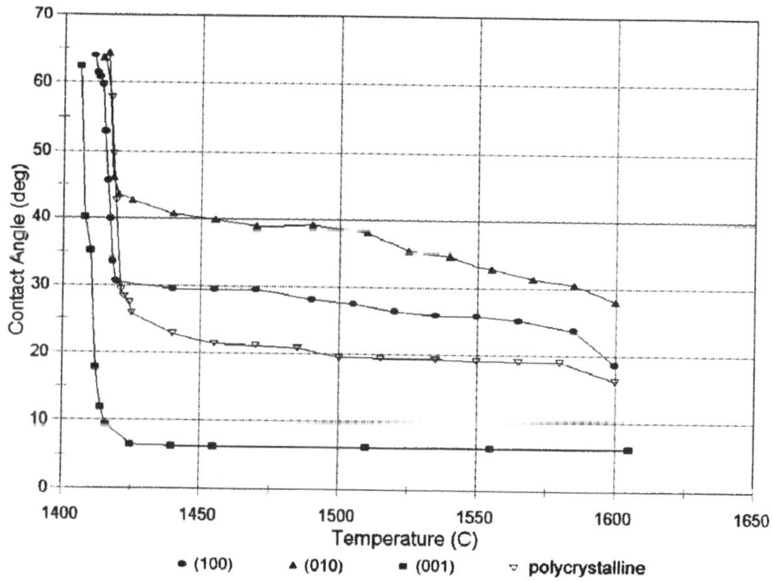

CONCLUSIONS

- Reactivity is related to substrate surface energy, γ_{SV}, which dominates the contact angle.

Substrate Orientation	Reactivity ↓	Borosilicate Glass		YAS-6 Glass	
		Contact angle	Moles of substrate reacted	Contact angle	Moles of substrate reacted
(010)		62	6.8	30	16
(100)		45	7	25	72
(001)		28	13.5	7	135

- The borosilicate glass reaction is simple dissolution of the 2:1 mullite substrate with oriented 2:1 mullite precipitates in the glass on cooling.
- The YAS reaction is complex and multi-phase.
- Both glasses penetrate grain boundaries of 3:2 polycrystalline mullite more deeply into the substrate while the borosilicate glass precipitates randomly oriented 2:1 mullite crystals on cooling.
- A new method of quantifying phase analysis from EDS qualitative elemental analysis in reactive wetting is shown which is in general agreement with observations.

REFERENCES

1. T.S. Huang, M.N. Rahaman, T. Mah, and T.A. Parthasarathay, "Anisotropic Grain Growth and Microstructural Evolution of Dense Mullite above 1550°C," J. Am. Ceram. Soc., **83** [1] 204-210 (2000).

2. T.S. Huang, M.N. Rahaman, B.T. Eldred, and P.D. Ownby, "Wetting of Mullite by Y_2O_3-Al_2O_3-SiO_2 and B_2O_3-SiO_2 Glasses," J. Mater. Res., **16** [11] 3223-3228 (2001).

3. B. Eldred. *Wetting of Single Crystal Mullite by Borosilicate and Yttrium-Aluminosilicate Glasses and Wetting Phenomena of Steels Containing Aluminum and Titanium.* Ph. D. Dissertation, University of Missouri – Rolla, Ceramic Engineering Department, 2004.

4. H. Schneider, K. Okada, and J. Pask, *Mullite and Mullite Ceramics.* J. Wiley, New York, 1994.

5. J. Chen, M.Y. Gu, and F.S. Pan, "Reactive Wetting of a Metal/Ceramic System," J. Mater. Res. **17** [4] 911-917 (2002).

6. B. Drevet, K. Landry, P. Vikner, and N. Eustathopoulos, "Influence of Substrate Orientation on Wetting Kinetics in Reactive Metal/Ceramic Systems," Scripta Mater. **35** [11] 1265-1270 (1996).

7. X.B. Zhou and J.T.M. DeHosson, "Reactive Wetting of Liquid Metals on Ceramic Substrates," Acta Mater. **44** [2] 421-426 (1996).

8. D.A. Weirauch, W.M. Balaba, and A.J. Perrotta, "Kinetics of the Reactive Spreading of Molten Aluminum on Ceramic Surfaces," J. Mater. Res. **10** [3] 640-650 (1995).

9. S. Kalogeropoulou, L. Baud, and N. Eustathopoulos, "Relationship Between Wettability and Reactivity in the Fe/SiC System," Acta Mater. **43** [3] 907-912 (1995).

Interfaces in High-Temperature Superconductors

CHEMICAL INTERACTIONS AND REACTION KINETICS OF THE $Ba_2YCu_3O_{6+x}/CeO_2$ SYSTEM

W. Wong-Ng, L.P. Cook, P. Schenck, I. Levin, Z. Yang, Q. Huang, and J. Frank
Materials Science and Engineering Laboratory
National Institute of Standards and Technology
Gaithersburg, MD 20899

ABSTRACT

Chemical interactions between the $Ba_2YCu_3O_{6+x}$ superconductor and the CeO_2 buffer layers employed in coated conductors have been modeled experimentally by investigating phase equilibria in the $Ba_2YCu_3O_{6+x}$ - CeO_2 system, and also by studying the kinetics of the reaction between $Ba_2YCu_3O_{6+x}$ and CeO_2. The $Ba_2YCu_3O_{6+x}$ - CeO_2 join within the BaO-Y_2O_3-CeO_2-CuO_x quaternary system is non-binary. At 810 °C and p_{O2} = 100 Pa, at a mole ratio of $Ba_2YCu_3O_{6+x}$: CeO_2 = 40 : 60, a phase boundary was found to separate two four-phase regions. On the $Ba_2YCu_3O_{6+x}$-rich side of the join, the four-phase region consists of $Ba_2YCu_3O_{6+x}$, $Ba(Ce_{1-z}Y_z)O_{3-x}$, BaY_2CuO_5, and Cu_2O; on the CeO_2 rich side, the four phases were determined to be $Ba(Ce_{1-z}Y_z)O_{3-x}$, BaY_2CuO_5, Cu_2O and CeO_2. There appears to be negligible solid solution formation of the types $Y_{1-z}Ce_zO_{3-x}$ and $Ce_{1-z}Y_zO_{2-x}$. The minimum melt temperature along the $Ba_2YCu_3O_{6+x}$ - CeO_2 join was determined to be ≈ 860 °C. The $Ba_2YCu_3O_{6+x}/CeO_2$ reaction is limited by solid-state diffusion, and the reaction kinetics obeys the parabolic rule, $x = Kt^{1/2}$ (where x is thickness of the reaction layer, t is time, and K is a constant related to the reaction rate); K was determined to be 3.6×10^{-2} μm/min$^{1/2}$ at 830 °C.

INTRODUCTION

With increased demand for electrical power, energy shortages and electricity outages have become common global problems. There are clear needs for improvement in electrical distribution grids and for more efficient utilization of energy resources. High-temperature superconductors have demonstrated potential for meeting these needs [1]. Consequently, there is an accelerated effort within the high T_c community on research and development of coated conductors for wire/tape applications [2-8]. These conductors are based on $Ba_2YCu_3O_x$ (Y-213) and $Ba_2RCu_3O_x$ (R=lanthanides) as the principal superconducting materials. They can be deposited on flexible metallic tapes, and the resulting materials show excellent current-carrying capability.

Specially prepared substrates form the basis for coated conductor fabrication. The two state-of-the-art technologies for producing biaxially-textured substrates are commonly known as ion beam assisted deposition (IBAD) [2-3], developed at Los Alamos National Laboratory, and rolling assisted bi-axially textured substrate (RABiTS) [4-8], developed at Oak Ridge National Laboratory. Typically, the architecture of a RABiTS film includes a number of layers of different materials deposited on the biaxially textured metallic substrate. These layers are the seed layer, barrier layer, lower cap layer, superconductor layer, and another cap layer on the superconductor [9]. The seed layer provides a thin epitaxial layer for protecting the substrate from oxidation during deposition of barrier layer. The barrier layer is a thick epitaxial layer for providing a physical/chemical barrier to substrate oxidation and substrate reaction with the superconductor layer. The lower cap layer provides additional protection for the superconductor film from chemical reaction, and also provides texture for crystallographic alignment. Examples of lower cap layer include CeO_2, $LaMnO_3$, $SrTiO_3$, Gd_3NbO_7, and $SrRuO_3$. The two most extensively used materials have been CeO_2 and $SrTiO_3$.

For a given combination of buffer layers that has been found to promote epitaxial growth of $Ba_2YCu_3O_{6+x}$, there may be unavoidable reactions at the interface between layers. Understanding of interfacial reactions of Y-213 phase with the buffer layers will provide information about how to avoid and/or control the formation of second phases. Phase equilibrium data will allow better interpretation of the results of TEM analysis of coated conductor interfaces.

This report describes the phase equilibria of the multi-component systems representing the interaction of $Ba_2YCu_3O_{6+x}$ with CeO_2 [10, 11], and also the kinetic study of phases formed at the interface. Phase equilibria and structural characterization studies were conducted by x-ray and neutron diffraction, and differential thermal analysis/thermal gravimetric analysis (DTA/TGA) [12]. Kinetic studies were conducted by depositing the Y-213 superconductor on polished CeO_2 pellets using pulsed laser deposition (PLD). X-ray diffraction and TEM were used for characterization of phase formation and to determine the thickness of the product layer.

EXPERIMENTAL[1]

Phase equilibria

(1) *Preparation of samples in the $Ba_2YCu_3O_{6+x}$-CeO_2 system*

For phase equilibra studies, a master batch of the superconductor phase $Ba_2YCu_3O_{6+x}$ was prepared by heating a mixture of BaO, Y_2O_3 and CuO under purified air (CO_2- and H_2O-scrubbed). The BaO starting material was produced from $BaCO_3$ (99.99 % purity, metals basis) by vacuum calcination in a vertical tube furnace. The following heating schedule was used: room temperature to 1300 °C in 20 h; isothermal at 1300 °C for 10 h; 1300 °C to room temperature in 20 h. To prepare $Ba_2YCu_3O_{6+x}$, stoichiometric amounts of BaO, Y_2O_3 and CuO were weighed out, well mixed and calcined in an atmospherically controlled high temperature furnace first at 850 °C, then at 930 °C repeatedly with intermediate grindings for about two weeks. Subsequently, thirteen samples with different ratios of $Ba_2YCu_3O_{6+x}$: CeO_2 were then prepared using the master $Ba_2YCu_3O_{6+x}$ batch and CeO_2 (5:95, 10:90, 20:80, 30:70, 35:65, 40:60, 45:55, 50:50, 60:40, 70:30, 80:20, 90:10, and 95:5). The processes of sample weighing, sample homogenization and pellet-pressing were performed inside a glove-box. Pelletized samples were placed inside individual MgO crucibles for annealing in a horizontal box-type controlled-atmosphere furnace. Transfer from the glove-box to the box furnace and vice versa was achieved via a second transfer vessel and an interlock system attached to the furnace.

(2) *Determination of the ternary phase diagrams*

Stoichiometric amounts of BaO, CeO_2, CuO and Y_2O_3 (99.99 % purity, metals basis) were first weighed inside a dry box according to the proper mole ratio, and were then mixed and pressed into pellets, followed by annealing in a controlled-atmosphere furnace. Sample transfer was accomplished by using an air-tight vessel. During the annealings, the oxygen pressure of Ar/O_2 mixtures was controlled using a mass flow meter and monitored at both the inlet and outlet of the furnace using a zirconia oxygen sensor. Samples were studied under atmospherically-controlled conditions at 810 °C with p_{O2} = 100 Pa (0.1 % O_2 by volume fraction) (in the absence

[1] Certain trade names and company products are mentioned in the text or identified in illustrations in order to adequately specify the experimental procedures and equipment used. In no case does such identification imply recommendation or endorsement by the National Institute of Standards and Technology.

of CO_2). Intermediate grindings and pelletizings took place until no further changes were detected in the powder x-ray diffraction patterns. Samples were processed for about 3 weeks each.

(3) *X-ray and neutron diffraction*

X-ray powder diffraction was used to identify the phases synthesized, to confirm phase purity, and to determine phase relationships. All x-ray patterns were measured using a hermetic cell designed for air-sensitive materials [13]. A computer-controlled automated diffractometer equipped with a θ-compensation slit and CuK_α radiation was operated at 45 kV and 40 mA. The radiation was detected by a scintillation counter and a solid-state amplifier. The Siemens software package and the reference x-ray diffraction patterns of the Powder Diffraction File (PDF) [14] were used for performing phase identification.

One selected sample within the $Ba(Ce_{1-z}Y_z)O_{3-x}$ solid solution (discussed later) was studied using neutron diffraction. Neutron diffraction data were collected with the 32 detector BT-1 diffractometer at the National Institute of Standards and Technology (NIST) Center for Neutron Research using a Cu(311) monochromator (λ=1.5396(1) Å). A 6 g sample was sealed into a 6 mm diameter vanadium container. Measurements were made at room temperature. The structural refinement was performed by using the GSAS software suite [15]. The structure of $BaCeO_3$ was employed as the initial model for the refinements [16].

(4) *Differential thermal analysis/Thermogravimetric Analysis (DTA/TGA)*

Simultaneous differential thermal analysis and thermogravimetric analysis (DTA/TGA) were used to study thermal events. Most experiments utilized primarily the DTA signal; the TGA signal was useful primarily in following oxygen gain/loss associated with the CuO_x component. DTA/TGA experiments were performed using an electronically upgraded Mettler TA-1 system fitted with an Anatech digital control and readout system. The DTA/TGA apparatus was calibrated against the α/β quartz transition (571 °C) and the melting point of NaCl (801 °C), and temperatures reported in this study have a standard uncertainty of ± 5 °C. Event temperatures were determined as the intersection of the baseline with the extrapolated linear portion of the rising DTA peak. Oxygen partial pressure during DTA/TGA was controlled using an analyzed Ar/O_2 mixtures. During the experiments, gas was continuously flowed through the sample region at a rate of 150 mL/min, and the oxygen pressure at the outlet of the DTA/TGA system was periodically checked with a zirconia sensor.

<u>Kinetic studies</u>

Kinetic studies were conducted by depositing the Y-213 superconductor on polished CeO_2 pellets (obtained from Alfa Aesar) using pulsed laser deposition (PLD).

(1) *Pulsed Laser Deposition Facility (Fig. 1)*

The laser deposition process is a high energy process with high deposition rates. Complex target compositions can be used, provided congruent vaporization occurs, with stoichiometric material transfer. The details of the experiments are as follows: excimer laser (248 nm, 100 mJ, and 10 Hz); deposition distance (3 cm); process gas (6.7 Pa oxygen); deposition rate (≈ 11.5 nm/min); and film thickness (≈ 570 nm).

(2) *Procedure*

To study the kinetics of reaction, a preliminary study to obtain the appropriate temperature range for heat treatments was first conducted. Seven films were prepared by PLD for this study, four for the preliminary studies to obtain the appropriate heat-treatment temperatures. Heat treatments were subsequently conducted in air at 790 °C, 810 °C, and 830 °C. For each sample, each subsequent cumulative heat treatment was in general doubled in time. X-ray diffraction was used to obtain intensity of selected reflections of these samples. The intensity values are assumed to be proportional to the amount of a particular phase, or the thickness of the reaction layer. Transmission electron microscopy (TEM) study was carried out to obtain microstructure, including identification of phases and thickness of reaction layer. From these observations, a kinetic model describing the progress of reaction could be determined.

Fig. 1. NIST Pulse Laser Deposition Facility

RESULTS AND DISCUSSION

Phase equilibria between $Ba_2YCu_3O_{6+x}$ and CeO_2

Figure 2 illustrates the equilibria along the $Ba_2YCu_3O_{6+x}$-CeO_2 join in the context of the BaO-½Y_2O_3-CuO_x-CeO_2 framework. It is seen clearly that two tetrahedral volumes (4-phase regions) corresponding to $Ba_2YCu_3O_{6+x}$-$BaCeO_3$-BaY_2CuO_5-Cu_2O and $BaCeO_3$-BaY_2CuO_5-Cu_2O-CeO_2 are mutually consistent and do not overlap. BaY_2CuO_5 is commonly known as the 'green phase'. These two tetrahedra share a common plane defined by BaY_2CuO_5, Cu_2O and $BaCeO_3$. In other words, as a composition vector passes through the two tetrahedra, only three phases are observed at the boundary. This phase boundary exists at the mole ratio of $Ba_2YCu_3O_{6+x}$: CeO_2 = 40 : 60. On the $Ba_2YCu_3O_{6+x}$-rich side, the four phases possible as a result of chemical interaction were found to be $Ba_2YCu_3O_{6+x}$, $BaCeO_3$, BaY_2CuO_5, and Cu_2O; whereas at the CeO_2 rich side, the four phases are $BaCeO_3$, BaY_2CuO_5, CeO_2 and Cu_2O. Apparently Cu_2O is more stable than CuO at p_{O2} = 100 Pa, in agreement with calculated CuO/Cu_2O equilibria [17].

Cerium is known to possess various oxidation states (+2, +3 and +4), therefore CeO_{2-x} undergoes a complex oxidation/reduction chemistry under different oxygen partial pressures. Under the conditions of the present study, CeO_2 is the only stable form of cerium oxide. The $BaCeO_3$ phase is a solid solution and is more appropriately written as $Ba(Ce_{1-z}Y_z)O_{3-x}$, although it is represented as a point compound on Fig. 2 as an approximation. Figure 3 gives the subsolidus diagram of the BaO-Y_2O_3-CeO_2 system that was determined using BaO as a starting component. Along the BaO-Y_2O_3 side, only two phases were found, namely, BaY_2O_4 and $Ba_3Y_4O_9$. There is no new ternary phase in this system except for the solid solution of the Y-doped $BaCeO_3$ phase, $Ba(Ce_{1-z}Y_z)O_{3-x}$. The range of this solid solution was determined to be rather small, namely, $0 \leq z \leq 0.13$. We have studied the structure of a solid solution member using neutron diffraction, which has essentially a perovskite-related structure. Despite extensive solid solution formed at higher temperature (for example, 20.5 % mole fraction of yttria in ceria at 1500 °C in air) [18-20], under ≈ 810 °C and p_{O2} = 100 Pa, we found no significant solid solution of between Y_2O_3 and CeO_2.

The structure of $Ba(Ce_{1-z}Y_z)O_{3-x}$ is of the perovskite type (Fig. 4). Y was found to substitute for the Ce site with an occupancy of 0.06(2), giving rise to the unit cell content of $Ba_4(Ce_{3.76}Y_{0.24})O_{11.36}$, or the chemical formula of $Ba(Ce_{0.94}Y_{0.06})O_{2.84}$, with Z=4, (space group of *Pmcn*, a = 8.7817(4) Å, b = 6.2360(4) Å, and c = 6.2190(3) Å, V = 340.57 Å^3, and a density 6.236 g/cm^3). The structure consists of corner-sharing [CeO_6] octahedral rotated around the a-, b-axes (10° to 11°), and the c-axis (1.2°). Ba atoms are situated in the interstices. As the sample was prepared under 0.1 % O_2, the oxygen content was determined to be less than the value of 3.0 from neutron diffraction study. A charge balance calculation leads to the chemical formula of $Ba(Ce^{3+}_{0.26}Ce^{4+}_{0.68}Y_{0.06})O_{2.84}$.

The phase diagrams of the CeO_2-Y_2O_3-CuO_x and the BaO-CeO_2-CuO_x systems are very simple, no ternary phase or solid solution was found in these systems under 0.1% O_2 and 810 °C. The details will be published elsewhere [21].

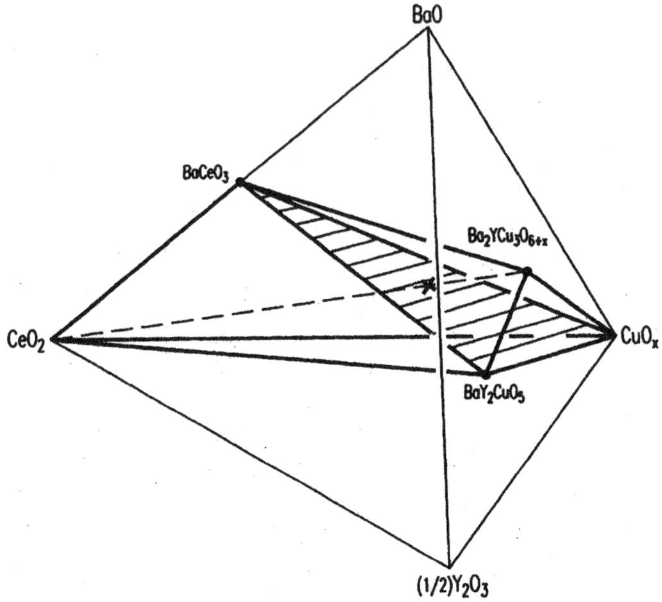

Fig. 2. The BaO-Y_2O_3-CuO_x-CeO_2 tetrahedron showing the two sub-volumes $Ba_2YCu_3O_{6+x}$-$Ba(Ce,Y)O_{3-x}$- BaY_2CuO_5- Cu_2O and $BaCeO_3$-BaY_2CuO_5- Cu_2O-CeO_2 within which the compositions of the $Ba_2YCu_3O_{6+x}$-CeO_2 join lie. "$BaCeO_3$ " is actually a solid solution of the formula $Ba(Ce_{1-z}Y_z)O_{3-x}$.

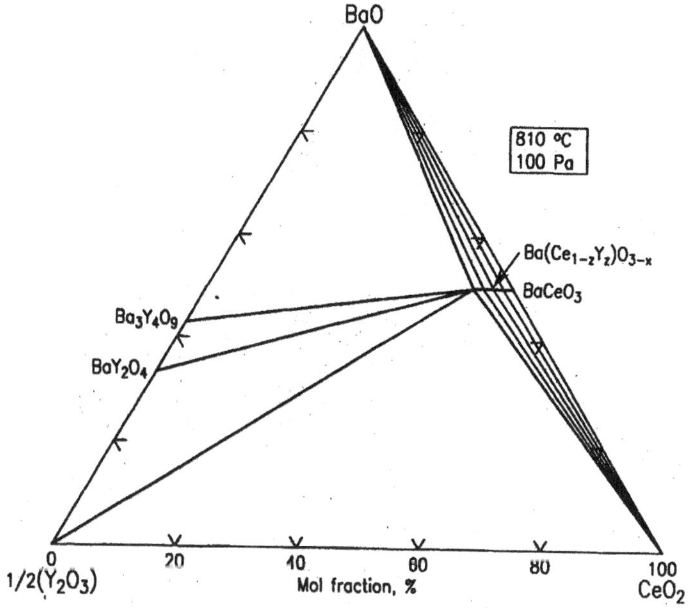

Fig. 3. Subsolidus diagram of the BaO-Y_2O_3-CeO_2 system determined at 810 °C, p_{O2}=100 Pa using BaO starting material.

Fig. 4. Crystal structure of $Ba(Ce_{0.94}Y_{0.06})O_{2.84}$ viewed along *c*-axis.

The initial melt temperature of the thirteen compositions can be categorized into two groups. Compositions of Y-213:CeO_2 ratio of 90:10, 80:20, 70:30, 60:40, and 45:55 gave an average melting temperature of 860 °C. Compositions with Y-213:CeO_2 ratio of 40:60, 30:70, 20:80 and 10:90 gave an average melting temperature of 959 °C. Therefore the minimum melting temperature of this non-binary join occurs near the $Ba_2YCu_3O_{6+x}$ side, and is estimated to take place according to the following reaction,

$$Ba_2YCu_3O_{6+x} + Cu_2O + BaY_2CuO_5 + Ba(Ce_{1-z}Y_z)O_{3-x} \rightarrow L \quad (\approx 860\ °C).$$

The next-lowest melting temperature event occurred at $\approx$ 959 °C (near the CeO_2 side). On the basis of these observations, since coated conductor processing takes place below 800 °C, the interaction of $Ba_2YCu_3O_{6+x}$ with CeO_2 will not cause melting during processing, unless significant surface energy effects exist.

Kinetic study of $Ba_2YCu_3O_{6+x}/CeO_2$

Figure 5 shows an optical image of two pellets, CeO_2 and CeO_2 coated with PLD $Ba_2YCu_3O_{6+x}$. The latter is typical of samples used in the kinetic studies.

Figure 6 shows the x-ray patterns of a sample heat-treated at 830 °C at various lengths (expressed in minutes) of heat treatment. One can follow the intensity development of peaks related to the products of $Ba(Ce_{1-z}Y_z)O_{3-x}$, BaY_2CuO_5, and reactant phase $Ba_2YCu_3O_{6+x}$. The decrease of the peak intensity of the reactant $Ba_2YCu_3O_{6+x}$ phase (by following reflection 031), and the increase of intensity of the products $Ba(Ce_{1-z}Y_z)O_{3-x}$ (reflection 213), and BaY_2CuO_5 (reflection 131) as a function of time, is evident. We observe a linear portion of the curves followed by an almost flat section at a longer duration time. A similar intensity trend was also observed for the 810 °C experiments. The increase of intensity of the peaks as a function of time again indicates the increased formation of $Ba(Ce_{1-z}Y_z)O_{3-x}$ and the green phase due to the reaction of $Ba_2YCu_3O_{6+x}$ with CeO_2.

In most ceramic reactions there is a reaction interface between reacting phases [22]. The overall process involves (1) transport of material to the interface, (2) reaction at the interface, and (3) sometimes transport of reaction products away from the interface. The two general classes of heterogeneous reactions are those controlled by transport rate, and those controlled by phase boundary reaction rate. For a reaction in which a compound is formed as the planar reaction layer and the rate of product formation is controlled by diffusion through the product layer, then the parabolic rate law, $x = Kt^{1/2}$, is observed, where 'x' is the thickness of the reaction layer, 't' is time, and K is a constant related to reaction rate (for the rare-limiting diffusion process).

We indeed found in all three temperatures that we studied, that the parabolic rate law was obeyed. The plots of intensity vs. $(time)^{1/2}$ for the experiments at 810 °C are shown in Fig. 7. For the products $Ba(Ce_{1-z}Y_z)O_{3-x}$ and BaY_2CuO_5, the curves are rather similar. There is the straight-line portion of these curves at a relatively short reaction time period, and then the change of

Fig. 5. Optical images of uncoated CeO_2 (left) and $Ba_2YCu_3O_{6+x}/CeO_2$ (right).

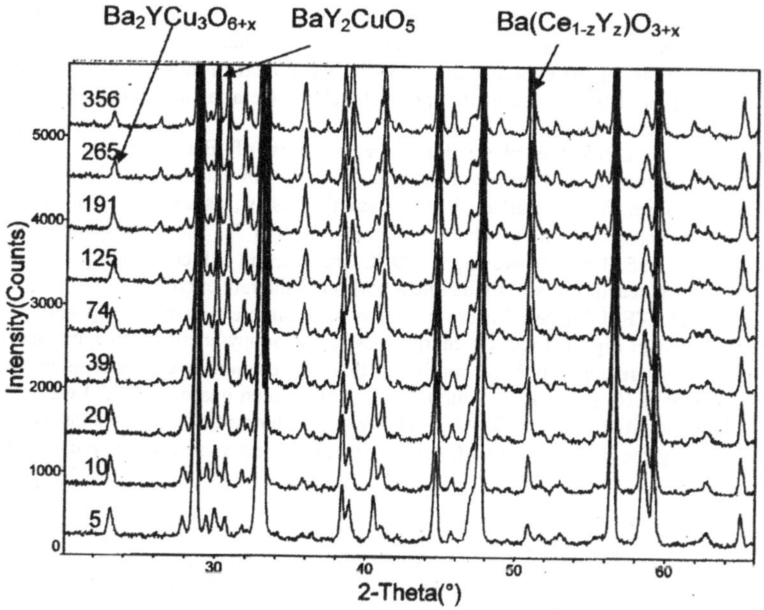

Fig. 6. Sequential x-ray diffraction patterns of $Ba_2YCu_3O_{6+x}/CeO_2$ pellet heat-treated at 830 °C for increasing cumulative time (minutes)

intensity as a function of time decreases dramatically. While the intensity of peaks corresponding to the reaction products increases, that of the reactant, namely the Y-213, decreases. The intensity vs. $t^{1/2}$ plots for the experiments at 830 °C are similar to those of 810 °C, namely, the straight-line portion of the curves for the products take place at a short range of reaction time, and then the slope becomes flat.

When one compares the phase formation features of $Ba(Ce_{1-z}Y_z)O_{3-x}$ at the three temperatures, different rates of formation are clear (Fig. 8). For example, at a higher temperature of 830 °C, the formation of $Ba(Ce_{1-z}Y_z)O_{3-x}$ was the fastest, as expected. The formation of $Ba(Ce_{1-z}Y_z)O_{3-x}$ was complete at about 180 min. At 810 °C the corresponding time was somewhat longer at 530 min. However, at 790 °C, the reaction still continues even at 3600 min. In other words, the reaction is still in the linear portion of the rate curve.

A bright-field TEM image of a cross-section of a sample heat-treated at 830 °C for 123 minutes is shown in Fig. 9. Electron energy loss spectroscopy/energy dispersive spectroscopy (EELS/EDS) measurements in TEM confirmed that the reaction layer, which is about 0.4 μm thick, consists of $Ba(Ce_{1-z}Y_z)O_{3-x}$. A mixture of Y-213, green phase, and CuO was observed in the immediate vicinity of $Ba(Ce_{1-z}Y_z)O_{3-x}$, whereas the outer part of the film consisted primarily of Y-213. Using the thickness of 0.40 μm, the parabolic equation gives a value of 3.6×10^{-2} $\mu m/min^{1/2}$ for K.

SUMMARY

We have studied the reaction of $Ba_2YCu_3O_{6+x}/CeO_2$ with respect to phase equilibria and kinetics. The reaction involves two 4-phase fields and the reaction products are $Ba(Ce_{1-z}Y_z)O_{3-x}$, BaY_2CuO_5, and CuO/Cu_2O. The minimum melting between $Ba_2YCu_3O_{6+x}$ and CeO_2 was determined to be 860 °C. Since coated conductor processing takes place below 800 °C, the interaction of $Ba_2YCu_3O_{6+x}$ with CeO_2 will not cause melting during processing, unless significant surface energy effects exist. We also followed the kinetics of reaction by monitoring the products $Ba(Ce_{1-z}Y_z)O_{3-x}$ and BaY_2CuO_5 as a function of time, and temperature. The reaction kinetics follows a simple parabolic rate law, characteristic of diffusion-limited processes. Further studies of interfacial reactions of $Ba_2YCu_3O_{6+x}$ with $SrTiO_3$ and with Gd_3NbO_7 are planned.

ACKNOWLEDGEMENTS

This project was partially supported by the U.S. Department of Energy. Nils Swanson of NIST is thanked for his graphical assistance. Mr. Edward Nie helped with the substrate preparation. We also acknowledge the valuable contributions from our colleagues, Drs. A. Goyal, M. Paranthaman, and R. Feenstra from Oak Ridge National laboratory, and Dr. T. Holesinger from Los Alamos National Laboratory.

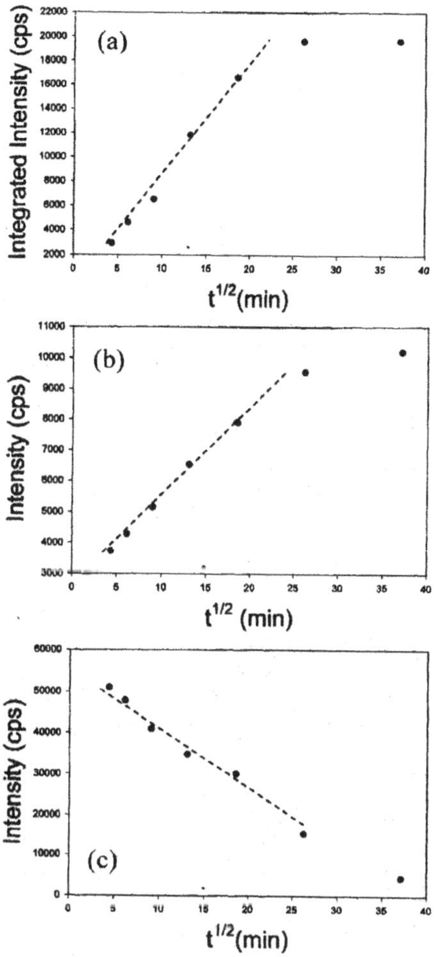

Fig. 7. Plots of intensity vs. (time, minutes)$^{1/2}$ (a) $Ba(Ce_{1-z}Y_z)O_{3-x}$, (213), (b) BaY_2CuO_5 (131), and (c) $Ba_2YCu_3O_{6+x}$ (031) for sample heat-treated at 810 °C. 'cps' stands for counts per second

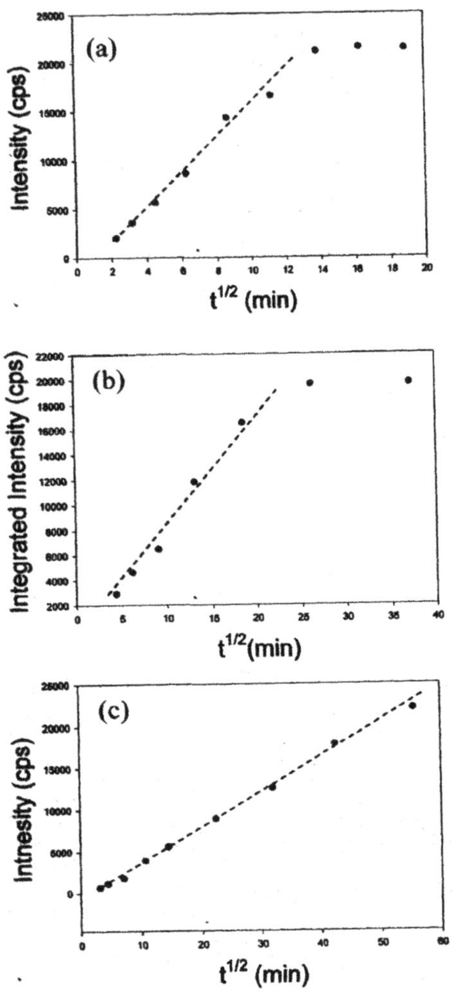

Fig. 8. Plots of intensity vs. (time, Minutes)$^{1/2}$ of $Ba(Ce_{1-z}Y_z)O_{3-x}$ (213) at three different temperatures, (a) 830 °C, (b) 810 °C, and (c) 790 °C. 'cps' stands for counts per second

Fig. 9. Bright-field TEM image of a cross-section of a sample heat-treated at 830 °C for 123 minutes

REFERENCES

[1] US Department of Energy high temperature superconductivity program; for details, visit web site http://www.eere.energy.gov/EE/power_superconductivity.html

[2] Foltyn S.R., Peterson E.J., Coulter J.Y., Arendt P.N., Jia Q.X., Dowden P.C., Maley M.P., Wu X.D., Peterson D.E., J. Mater. Res. **12** 2941-2946 (1997).

[3] Iijima Y., Hosaka M., Tanabe N., Sadakata N., Saitoh T., Kohno O., Takeda K., J. Mater. Res. **12** 2913-2923 (1997).

[4] Malozemoff A.P., Annavaraou S., Fritzemeier L., Li Q., Prunier V., Rupich M., Thieme C., Zhang W., Goyal A., Paranthaman M., and Lee D.F., Supercond. Sci. Technol. **13** 473-476 (2002).

[5] Paranthaman M, Park C., Cui X, Goyal A., Lee D.F., Martin P.M., Chirayil T.G., Verebelyi D.T., Norton D.P., Christen D.K., and Kroeger D.M., J. Mater. Res **15** (12) 2647-2652 (2000).

[6] Goyal A., Lee D.F., List F.A., Specht E.D., Feenstra R., Paranthaman M., Cui X., Lu S.W., Martin P.M., Kroeger D.M., Christen D.K., Kang B.W., Norton D.P., Park C., Verebelyi D.T., Thompso J.R., Williams R.K., Aytug T., and Cantoni C., Physica C **357** 903-913 (2001).

[7] Aytug T., Goyal A., Rutter N., Paranthaman M., Thompson J.R., Zhai H.Y., and Christen D.K., J. Mater. Res. **18 [4]** 872-877 (2003).

[8] Paranthaman M., Goyal A., List F.A., Specht E.D., Lee D.F., Martin P.M., He Q., Christen D.K., Norton D.P., Budai J.D., Kroeger D.M., Physica C **275** 266-272 (1997).

[9] Rupich M., American Superconductor Corp, private communication.

[10] Wu L., Zhu Y., Solovyov V.F., Wiesmann H.J., Moodenbaugh A.R., Sabatini R.L., Suenaga M., J. Mater. Res. **16** [10] 2869 – 2884 (2001).

[11] Holesinger T.G., Gibbons B.J., Coulter J.Y., Foltyn S.R., Groves J.R., and Arendt P.N., Materials Research Society Symposium Proceedings Volume 689, *Materials for High-Temperature Superconductor Technologies*, Ed. M.P. Paranthaman, M.W. Rupich, K. Salama, J. Mannhart, and T. Hasegawa, Symposium held November 26-29, 2001, Boston, MA, 2002, pp. 329-336.

[12] Wong-Ng W., Cook L.P., Ceramic Transaction, 18, Superconductivity in Ceramic Superconductors 2, Nair K.M., Balachandran U., Chiang Y-M., Bhalla A., eds., Am. Ceram. Soc., Westerville, OH, 1991, pp. 73.

[13] Ritter J. J., Powd. Diffr. **3** 30-31 (1988).

[14] PDF, Powder Diffraction File, produced by International Centre of Diffraction Data (ICDD), 12 Campus Blvd., Newtown Squares, PA. 19073-3273.

[15] Larson A.C., von Dreele R.B. (1992). *GSAS-General Structure Analysis System*, US Government contract (W-7405-ENG-36) by the Los Alamos National Laboratory, which is operated by the University of California for the U.S. Department of Energy.

[16] Guha J.P., Kolar D., J. Mater. Sci. **6** (9) 1174-1177 (1971).

[17] Chase, Jr. M.W., NIST-JANAF Thermochemical Tables, 4th Edition, J. Phys. Chem. Ref. Data, Monograph No 9, 1998, pp.1952.

[18] Wang D.Y., Park D.S., Griffith J., Nowick A.S., Solid State Ionics **2** 95-105 (1981).

[19] Balazs G.B., Glass R.S., Solid State Ionics **76** 155-162 (1995).

[20] Longo V., Podda L., J. Mater. Sci. **16**(3) 839-841 (1981).

[21] Wong-Ng W., Yang Z., Cook L.P., Huang Q., Kaduk J.A.[1], and Frank J., Solid State Sciences, in print.

[22] Kingery W.D., Bower H.K., and Uhlmann D.R., *Introduction to Ceramics*, 2nd Ed., pp. 381-447, John Wiley & Sons, New York, USA, 1976.

SUPERCONDUCTING PROPERTIES OF $(Y_{1-x}Gd_x)Ba_2Cu_3O_{7-\delta}$ COMPOSITES PROCESSED IN PARTIAL OXYGEN ATMOSPHERES

T. J. Haugan, J. M. Evans, I. Maartense, P. N. Barnes
Air Force Research Laboratory
1950 Fifth St. Bldg 450
Wright-Patterson AFB, OH 45433-7251

ABSTRACT

The effect of Gd substitution for Y on the superconducting properties of $(Y_{1-x}Gd_x)Ba_2Cu_3O_{7-\delta}$ {(Y,Gd)123} composites was studied in powders processed by solid-state reaction and equilibrated at 910°C in air and finally in 1% O_2 atmosphere. Powders were subsequently annealed in 100% O_2 atmosphere at temperatures < 500 °C to maximize T_c. After processing, the powders were nearly pure single-phase as determined by X-ray diffraction. With increasing Gd substitution, the T_cs as measured by ac susceptibility showed a steady increase from 92 K to 95 K. The flux pinning $J_c(H)$ properties of (Y,Gd)123 compared to Y123 at 65 K and 77 K and magnetic fields up to 9T were consistently increased up to 10 times with Gd substitution, but only the Gd = 0.6 composition showed an increase compared to Gd123 for H > 7 T. A "fish-tail" effect was measured for the Gd-substituted powders at 65K, with $H_{peak} \sim 3.5$ T and ~ 6 T for Gd = 0.2 and Gd = 0.6 compositions, respectively.

INTRODUCTION

The high critical transition temperature (Rare-Earth)$Ba_2Cu_3O_{7-z}$ (RE123) superconductors are being considered for applications including thin film coated conductors and bulk devices because of their high superconducting transition temperatures (T_c) > 90 K, and enhanced critical current density (J_c) at 77 K in application-relevant magnetic fields. While these materials have desirable qualities at 65K or 77 K, it is of interest to increase the $J_c(H)$ properties even further by increasing the magnetic flux pinning properties. Different methods can be considered to introduce flux pinning defects into the superconductors, including irradiation and addition of second-phase defects or precipitates. One method being considered by many groups is to substitute rare-earth cations into the RE123 compound [1-37]. Different variations of this theme have been studied, including substitution in (Y,RE)123 with RE = Ho, Dy, Gd, Eu, and Pr, and various other combinations of rare-earths such as mixed (RE_1,RE_2)123 or (RE_1,RE_2,RE_3)123 compositions such as (Nd,Eu,Gd)123 [1-37]. Recent studies have demonstrated that significant increases of $J_c(H)$ can be achieved in mixed RE123 materials and particularly for RE ~ 0.33 or ~0.66 in (Y,RE)123 compositions; e.g. for RE = Ho, Gd, Dy, Sm, and Eu [19,32-35,26], and for mixed $(Nd_{0.333},Eu_{0.333},Gd_{0.333})$123 [7,26]. Processing and properties of chemically substituted YBCO have been reviewed in several articles [23-26], and RE substitutions have been summarized recently for flux pinning [26]. At least 150 papers have been published on (Y,RE)123 flux pinning, however there tends to be limited systematic studies on $(Y_{1-x}RE_x)$123 for the complete range of x = 0 – 1 [26]. Figure 1 plots the increase of $J_c(H)$ from several of these studies, and indicates how different effects are measured depending on the RE ion substitution.

Possible mechanisms by which RE substitutions can increase flux pinning include: **(a)** addition of second-phase defects by precipitation or composition changes, **(b)** formation of finely distributed lower T_c components from the mixed solubility of RE with Ba and intersolubility of RE or other mechanisms, **(c)** randomly distributed oxygen-deficient zones which have lower T_cs [7,8], or **(d)** intrinsic stresses resulting from the lattice mismatched RE123 agglomerations [21,22]. The finely distributed lower T_c components are suggested as a cause of the so-called "fish-tail" effect, where as the magnetic field is increased, the J_c temporarily increases as the lower T_c components transition to normal behavior after which the J_c decreases at much higher applied magnetic fields [7,8,29]. The peak of the J_c maximum in these materials typically occurs at applied fields of about 2 T to 3 T, and the "fish-tail" peaks been observed in melt-processed (Y,Gd)123 [29]. The distribution and size of the RE123 unit cell agglomerations is not precisely known, however presumably would variably affect the T_c and $J_c(H)$ properties.

Studies of the system $(Y_{1-x}Gd_x)Ba_2Cu_3O_{7-\delta}$ have, to our knowledge, been limited thus far to melt-processed or powder materials [27-31] or thin films [32,33]. In the work herein, the effect of composition changes in (Y,Gd)123 was studied in solid-state powders annealed in 1% O_2 atmosphere to achieve chemical equilibrium, similar to previous work on $(Y_{1-x}Nd_x)Ba_2Cu_3O_{7-\delta}$ powders [16,17]. Powder compositions are expected to have different properties than melt-processed materials or thin films, where non-equilibrium processes can affect the physical properties. In melt-processed materials it is virtually impossible to eliminate the formation of large size and volume-fraction second-phase defects such as RE211, which can affect pinning. With powder processing, it is possible to significantly minimize or eliminate large second-phase defects, and investigate other causes of pinning more related to the 'intrinsic' or nanosize properties of the processed material.

For the studies herein, powders were annealed in 1% oxygen atmosphere, which is expected to enhance site occupation of Gd on the RE rather than the Ba site [15,18]. Similar to the work on Nd, this is expected to sharpen and increase the T_c transition, however it could decrease flux pinning by reducing the formation of lower T_c solid-solutions [8,15-18]. Previously we have studied the effect of substitution of Nd for Y studies on powders in the (Y,RE)123 system [16,17]. In studying RE = Nd substitution, the T_c in particular was noted to be quite low for some of the Nd-rich and Ba-poor compositions after annealing in air [16,17]. Subsequent processing in reduced oxygen atmospheres sharpened and increased the T_c transition of the (Y,Nd)123 compositions and improved the superconducting properties, as expected [5,15,18]. Since Gd is quite close to Nd in ionic size when compared to other RE elements, it was decided to test similar conditions by processing all (Y,Gd)123 powders with final high-temperature annealing in 1% O_2 atmosphere. How effective this is for the mixed compositions is demonstrated herein.

In this paper, only solid-solution single-phase compositions are considered that preserve the single-crystal structures yet perturb the superconductor in localized regions surrounding the chemically substituted volumes. The single-phase regions for (Y,Gd)123 with Ba substitution were measured by Wong-Ng et al [37], as shown in Figure 2. The single-phase region for $(Y,Gd)_{1+x}Ba_{2-x}Cu_3O_{7-y}$ is slightly smaller compared to $(Y,Nd)_{1+x}Ba_{2-x}Cu_3O_{7-y}$ because of the smaller ionic size of Gd^{+3} compared to Nd^{+3}, which makes it more difficult for Ba site substitution [16,17,37]. It's expected that with Gd substitution on the Y or Ba sites, the defect density of the locally-perturbed areas can be high enough to enhance pinning if only 5% or less of the defects are effective pinning centers [26]. Whether the (Y,Gd)123 agglomerations are effective for pinning are studied herein.

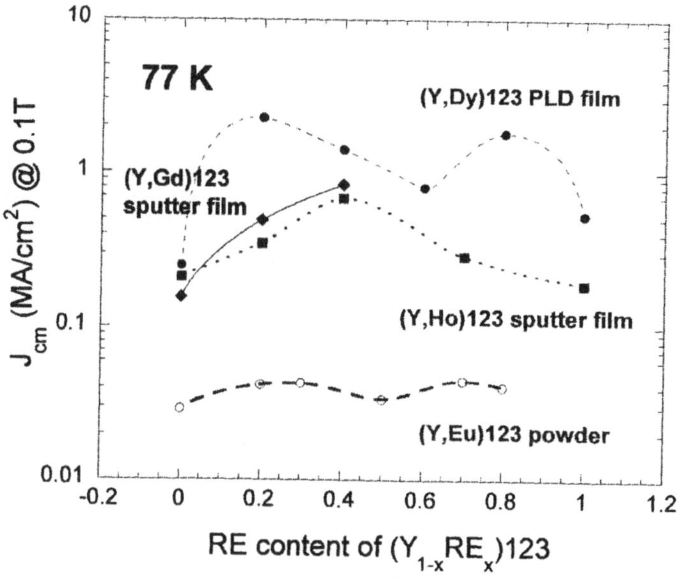

Figure 1. J_c by magnetization methods (J_{cm}) vs RE substitution for various (Y,RE)123 materials processed by various methods: RE = Eu [36], Ho [34], Gd [33] and Dy [35].

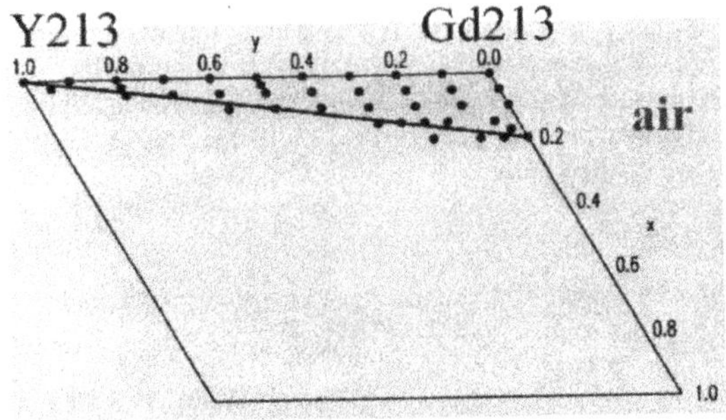

Figure 2. Compositions of $(Y_{1+x-y}Gd_y)Ba_{2-x}Cu_3O_{7-\delta}$ powders processed in air that were nominal single-phase are shown in the darker shaded phase spaces [from Wong-Ng et al, Ref. 37]. Note that the designation RE213 in these diagrams referes to the $(RE)_1Ba_2Cu_3O_{7-\delta}$ structure which is designated as RE123 in this paper.

EXPERIMENTAL**

Experimental methods for this work use procedures identical to previous work for (Y,Nd)123 powders [16,17]. Superconducting powders were prepared by the solid-state reaction method, using starting reactants of Gd_2O_3, Y_2O_3, $BaCO_3$, and CuO ($\geq$ 99.95% purity). The powders were dehydrated at 450 °C prior to weighing [15,18]. The powders were mixed and ground with mortar and pestle, calcined by slow heating 650 °C to 850 °C at 25 °C/h, and subsequently annealed with intermediate grinding steps at 880 °C and 910 °C. Powders were annealed at 910 °C with intermediate grinding until phase equilibrium was reached (3-4 annealings), as determined by X-ray diffraction (XRD). The powders were reacted in ~1 cm diameter pellets (0.5 g to 1 g batches), formed by lightly pressing at ~5-10x10^6 Pa in molds. X-ray diffraction was performed with a Rigaku diffractometer. A step size of 0.03° was used for the θ-2θ scans.

After single-phase compositions were obtained, an additional processing step in an 1% O_2 atmosphere was performed; the annealing temperature was at 820°C to 920°C increasing in 20°C increments as the Gd content increased from 0 to 1.0, and pellets were reacted on mixed (Gd,Y)211 pellets to minimize reactions with the substrate. After annealing in 1% O_2 atmosphere, the samples were checked with XRD to determine if the single phase composition was preserved or whether some slight reaction (melting or decomposition) might have occurred to cause formation of Y_2BaCuO_5 and $BaCuO_2$ phases (for example). A final annealing step was performed in 100% O_2 atmosphere at 460°C to 260°C, decreasing in 40°C increments as the Gd content increased from 0 to 1.0; this assumed a linear gradient in temperature between the optimized annealing temperatures for Y123 and Gd123 compositions [38]. The optimal annealing temperature in 100% O_2 atmosphere for the mixed RE123 compounds to maximize the oxygen content and T_c is yet unknown, therefore this approximation was used.

Superconducting properties of powders were measured with a vibrating sample magnetometer (VSM - Quantum Design, PPMS). Magnetization-applied field (M-H) hysteresis loops at different temperatures were made by heating samples to 100 K and zero-field cooling (ZFC) to the measurement temperature. The magnetic field was increased using a constant ramp of ~ 8000 (A/m•sec), and complete M-H loops were measured from 0 T to + 9T to – 9T to + 9T and back to 0 T. The magnetic J_c was estimated using the extended Bean critical current model $J_c = 15(\Delta M)/R$ where ΔM is the volume magnetization, and R is the radius of the separated percolating superconducting grains which was roughly approximated as 0.00005 cm [39].

The superconducting transition temperature (T_c) was also measured using an AC susceptibility technique with the amplitude of the magnetic sensing field, h, varied from 0.025 Oe to 2.2 Oe, at a frequency of approximately 4 kHz. The AC susceptibility technique provides information about primary and secondary transitions of the bulk samples. T_cs were measured on palletized samples with very small thickness. Samples were mounted onto the end of a sapphire rod and measured as the samples were warmed through the transition region at very slow rate of ~ 0.06 K/min. The T_c measurements were accurate within $\leq$ 0.1 K at three calibration points: liquid He at 4.2 K, liquid N_2 at 77.2 K, and room temperature. Different methods of measuring powders were tested, such as mounting pellets and powders placed in silicone grease, were used with the different methods giving the same T_c values within about 0.1 K.

RESULTS

X-ray diffraction measurements of the mixed (Y,Gd)123 compositions indicated single-phase RE123 structures were achieved with Gd substitution, similar to measurements for powders processed in air shown in Figure 2 [Wong-Ng et al., Ref. 37]. A small amount of RE211 and $BaCuO_2$ second-phase materials were present for the Gd = 0.2 composition, however the volume fraction was estimated as < 5% from XRD peak intensities. All other compositions were virtually phase-pure RE123 as measured by XRD. The phase purity of the (Y,Gd)123 powder compositions in Figure 2 were similar to phase purities of (Y,Nd)123 powders processed by similar methods [16,17].

The superconducting transition onsets of the $(Y_{1-x}Gd_x)$123 compositions reacted in air followed by 1% O_2 final processing temperature are shown in Figure 3, and compared to previous results obtained for $(Y_{1-x}Nd_x)$123 processed by the exact conditions [17]. As seen in Figure 3, the onset T_cs showed a steady increase with Gd substitution except for a small decrease at the Gd = 0.2 composition for which the RE211 phase inclusions were observed. Interestingly, a similar decrease at Gd = 0.2 was observed by other authors [40], consistent with the results in Figure 3. Comparing results of (Y,Gd)123 with (Y,Nd)123 in Figure 3, the onset T_c onset transitions showed similar trends however the T_c increased significantly for Gd > 0.6 and increased slightly in general for RE < 0.6 substitution.

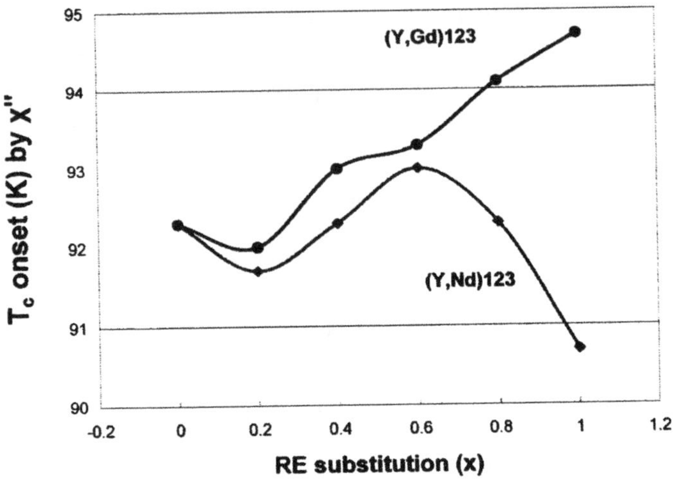

Figure 3. Transition temperature of the bulk superconducting volume fraction of $(Y_{1-x}Gd_x)$123 powders processed in air followed by 1% O_2, compared to $(Y_{1-x}Nd_x)$123 processed by similar methods [16,17].

Figure 4 plots the $J_c(H)$ properties of the powders for varying Gd substitution at 65 K and 77 K. In general, the J_cs increased with increasing Gd substitution similar to the increase of T_c shown in Figure 3. The Gd substituted powders had significant increases of $J_c(H)$ compared to Y123; however, compared to the pure Gd123 composition only Gd = 0.6 had an increase of $J_c(H)$ for H > 7T at 77 K. These consistent results suggest that the Gd substituted compositions are of interest, for example for further testing in mixed composition $(Y_{1-x}Gd_x)$123 thin films. Preliminary results with Gd123 thin films deposited at lower temperatures indicate that significant increases of $J_c(H)$ can be achieved compared to Y123, which are consistent with the results in Figure 4 showing enhancement of Gd123 compared to Y123 [3].

Figure 4 also indicates that several powders exhibited the "fish-tail" effect, especially at 65K for Gd = 0.2 and Gd = 0.6 compositions. The "fish-tail" effect is when as the magnetic field is increased the J_c temporarily increases as the lower T_c components transition to normal behavior, after which the J_c decreases at much higher applied magnetic fields [7,8,29]. Interestingly, the peak of the $J_c(H)$ curves shifted for different compositions; e.g. $H_{peak} \sim 3.5$ T for Gd = 0.2 and $H_{peak} \sim 6$ T for Gd = 0.6. The results for Gd = 0.6 are somewhat different from results of other authors studying melt-processed material where $H_{peak} \sim 2.3$ T at 65 K was measured [29].

To better measure the pinning improvements, Figure 5 plots the increase of $J_c(H)$ from Figure 4 at select magnetic fields as a function of Gd substitution. Compared to Y123, the Gd substituted powders had increases at 77K about 2-3 times at $H \leq 0.2$ T (low-fields) and about 7-10 times at $\geq$ 1T (high-fields). The increases at 65K were different however, with more moderate increases of ~ 3-7 times observed varying with the H field. Compared to Gd123, however only the Gd = 0.6 composition showed similar or slight improvements of $J_c(H)$ at 77 K.

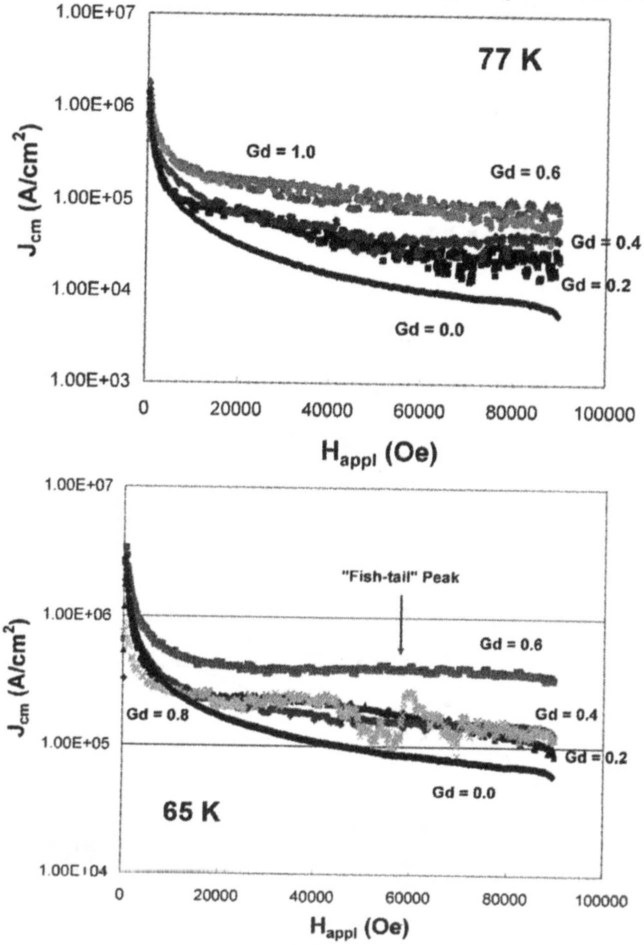

Figure 4. Critical current density by VSM (J_{cm}) as a function of applied magnetic field (H_{appl}) at 77 K (upper) and 65K (lower) of $(Y_{1-x}Gd_x)123$ powders with final high-temperature annealing in 1% O_2 atmosphere.

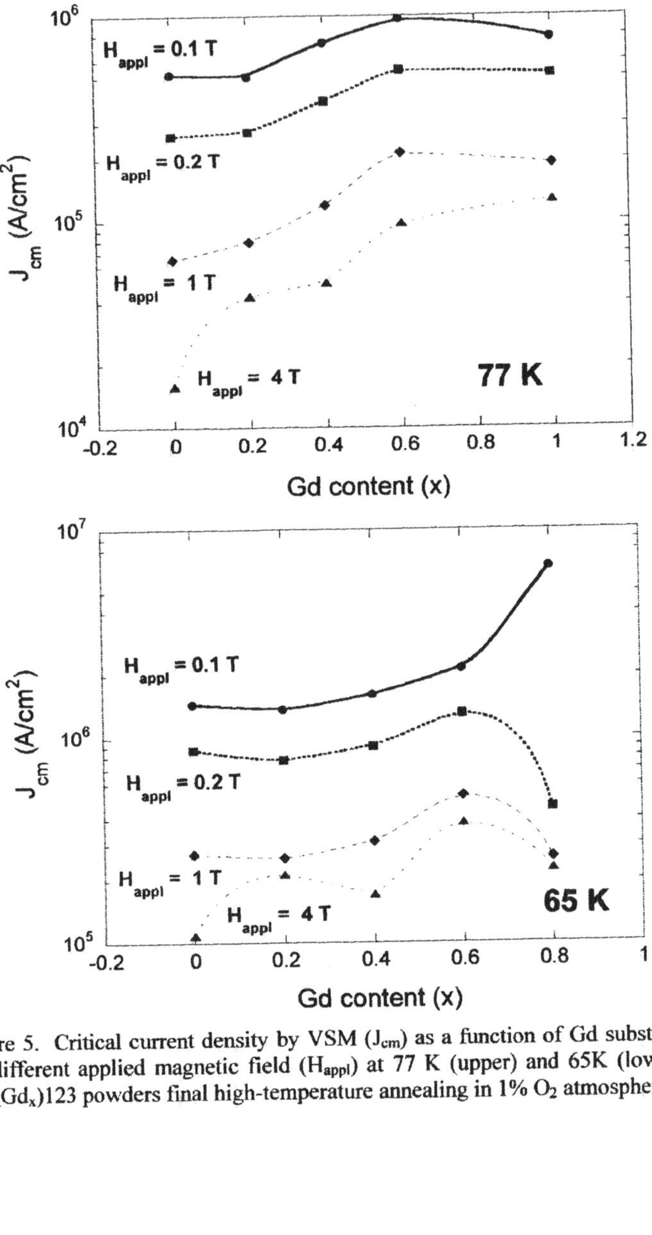

Figure 5. Critical current density by VSM (J_{cm}) as a function of Gd substitution for different applied magnetic field (H_{appl}) at 77 K (upper) and 65K (lower) of $(Y_{1-x}Gd_x)123$ powders final high-temperature annealing in 1% O_2 atmosphere.

CONCLUSIONS

The flux pinning and superconducting properties of $(Y_{1-x}Gd_x)Ba_2Cu_3O_{7-\delta}$ powders were studied for Gd substitution = 0 to 1 and with final high-temperature annealing in 1% O_2 atmosphere. Processing was found to form single-phase structures, and the superconductivity and T_c and J_c were significantly increased with increasing Gd substitution. A steady increase of T_c up to 95 K was measured with Gd substitution except for a slight decrease at Gd = 0.2. The $J_c(H)$ increased substantially up to 10 times for all mixed (Y,Gd)123 compositions compared to Y123 for a wide range of applied magnetic fields, however only one composition Gd = 0.6 showed improvement of $J_c(H)$ compared to Gd123 at 77K for limited magnetic fields > 7 T.

A distinct "fish-tail" effect was observed at 65K for Gd = 0.2 and 0.6 compositions with $J_c(H)$ increasing at moderate fields of 3-6 T compared to fields of 1-2 T, similar to previous published results for Gd = 0.6 [29]. H_{peak} values at 65K of ~3.5 T and ~ 6 T were achieved for Gd = 0.2 and Gd = 0.6 compositions, respectively.

ACKNOWLEDGEMENTS

The authors would like to thank L. Brunke and J. Kell for assistance with VSM measurements.

** Certain commercial equipment, instruments, or materials are identified in this paper in order to specify the experimental procedure adequately. Such identification is not intended to imply recommendation or endorsement by the Air Force Research Laboratory, nor is it intended to imply that the materials or equipment identified are necessarily the best available for the purpose.

REFERENCES

[1]Y. Feng, A. K. Pradhan, Y. Zhao, Y. Wu, N. Koshizuka, and L. Zhou, "Influence of Ho substitution for Y on flux pinning in melt-processed YBCO superconductors," *Physica C*, **357-360**,799-802 (2001).

[2]V. Selvamanickam, Y. Xie, J. Reeves, and Y. Chen, "MOCVD-Based YBCO-Coated Conductors," *MRS Bulletin*, **August 2004**, 579-582 (2004).

[3]K. Matsumoto, D. takahara, T. Horide, A. Ichinose, S. Horii, Y. Yoshida, M. Mukaida, and K. Osamura, "High-J_c Gd-Ba-Cu-O Epitaxial Films Prepared by Pulsed Laser Deposition," *IEEE Trans. Appl. Supercond.*, **15(2)**, 2719-2722 (2005).

[4].G. Osabe, T. Takizawa, S.I. Yoo, N. Sakai, T. Higuchi, M. Murakami, "Studies of the $Nd_{1+z}Ba_{2-x}Cu_3O_y$ solid solutions," *Mat. Sci. Engineer.*, **B65**, 11-16 (1999).

[5]H. H. Wen, Z. X. Zhao, Y. G. Xiao, B. Yin, and J. W. Li, "Evidence for flux pinning induced by spatial fluctuation of transition temperatures in single domain $(Y_{1-x}Pr_x)Ba_2Cu_3O_{7-\delta}$ samples," *Physica C*, **251**, 371-378 (1995).

[6]E. S. Reddy, P. V. Patanjali, E. V. Sampathkumaran, R. Pinto, "Fabrication and superconducting properties of ternary $REBa_2Cu_3O_y$ thin films," *Physica C*, **366**, 123-128 (2002).

[7]M. R. Koblischka, M. Muralidhar, M. Murakami, "Flux pinning sites in melt-processed $(Nd_{0.33}Eu_{0.33}Gd_{0.33})Ba_2Cu_3O_y$ superconductors," *Physica C*, **337**, 31-38 (2000).

[8]M. Jirsa, M. R. Koblischka, T. Higuchi, M. Muralidhar, M. Murakami, "Comparison of different approaches to modelling the fishtail shape in RE-123 bulk superconductors," *Physica C*, **338**, 235-245 (2000).

[9]C. Varanasi, P. J. McGinn, H. A. Blackstead, and D. B. Pulling, "Nd Substitution in Y/Ba Sites in Melt Processed $YBa_2Cu_3O_{7-\delta}$ Through Nd_2O_3 Additions," *Journal of Electronic Materials*, **24[12]**, 1949-1953 (1995).

[10]D. N. Matthews, J. W. Cochrane, G. J. Russell, "Melt-textured growth and characterization of a $(Nd/Y)Ba_2Cu_3O_{7-\delta}$ composite superconductor with very high critical current density," *Physica C*, **249**, 255-261 (1995).

[11]P. Schätzle, W. Bieger, U. Wiesner, P. Verges and G. Krabbes, "Melt Processing of (Nd,Y)BaCuO and (Sm,Y)BaCuO composites," *Supercond. Sci. and Technol.*, **9**, 869-874 (1996).

[12]A. S. Mahmoud and G. J. Russell, "Large crystals of the composite Y/Nd(123) containing various dopants grown by melt-processing in air," *Supercond. Sci. and Technol.*, **11**, 1036-1040 (1998).

[13]X. Yao, E. Goodilin, Y. Yamada, H. Sato and Y. Shiohara, "Crystal growth and superconductivity of $Y_{1-x}Nd_xBa_2Cu_3O_{7-\delta}$ solid solutions," *Applied Superconductivity*, **6[2-5]**, 175-183 (1998).

[14]D. K. Aswal, T. Mori, Y. Hayakawa, M. Kumagawa, "Growth of $Y_{1-z}Nd_zBa_2Cu_3O_x$ single crystals," *Journal of Crystal Growth*, **208**, 350-356 (2000).

[15]H. Wu, K. W. Dennis, M. J. Kramer, and R. W. McCallum, "Solubility Limits of $LRE_{1+x}Ba_{2-x}Cu_3O_{7+\delta}$," *Applied Superconductivity*, **6[2-5]**, 87-107 (1998).

[16] T. J. Haugan, M. E. Fowler, J. C. Tolliver, P. N. Barnes, W. Wong-Ng, L. P. Cook, "Flux Pinning and Properties of Solid-Solution $(Y,Nd)_{1+x}Ba_{2-x}Cu_3O_{7-\delta}$ Superconductors", in *Processing of High Temperature Superconductors*, Ceramic Transactions **Vol. 104**, edited by A. Goyal, W. Wong-Ng, M. Murakami, J. Driscoll (American Ceramic Society, Westerville OH, 2003), p. 299 - 307.

[17]T.J. Haugan, J.M. Evans, J.C. Tolliver, I. Maartense, P.N. Barnes, W. Wong-Ng, L.P. Cook, R.D. Shull, "Flux Pinning and Properties of Solid-Solution $(Y,Nd)_{1+x}Ba_{2-x}Cu_3O_{7-d}$ Superconductors Processed in Air and Partial Oxygen Atmospheres" in *Fabrication of Long-Length and Bulk High Temperature Superconductors,*, Ceramic Transactions **Vol. 149**, edited by R. Meng, A. Goyal, W. Wong-Ng, H.C. Freyhardt, and K. Matsumoto (American Ceramic Society, Westerville OH, 2004), p. 151 – 162.

[18]R. W. McCallum, M. J. Kramer, K. W. Dennis, M. Park, H. Wu, and R. Hofer, "Understanding the Phase Relations and Cation Disorder in $LRE_{1+x}Ba_{2-x}Cu_3O_{7+\delta}$," *J. of Electr. Mater.*, **24[12]**, 1931-1935 (1995).

[19]J. L. MacManus-Driscoll, S. R. Foltyn, B. Maiorov, Q. X. Jia, H. Wang, A. Serquis, L. Civale, Y. Lin, M. E. Hawley, M. P. Maley, D. E. Peterson, "Rare earth ion size effects and enhanced critical current densities in $Y_{2/3}Sm_{1/3}Ba_2Cu_3O_{7-x}$," *Appl. Phys. Lett.* **86**, 032505-032507 (2005).

[20]J. L. MacManus-Driscoll, S. R. Foltyn, Q. X. Jia, H. Wang, A. Serquis, B. Maiorov, L. Civale, Y. Lin, M. E. Hawley, M. P. Maley, D. E. Peterson, "Systematic enhancement of in-field

critical current density with rare-earth ion size variance in superconducting rare-earth barium cuprate films," *Appl. Phys. Lett.*, **84**, 5329-5331 (2004).

[21]Y. Li, Z.-X. Zhao, "Stress-field pinning induced by the lattice mismatch in 123 phase," *Physica C*, **351** , 1-4 (2001).

[22]Y. Li, L. Cui, G. Cao, Q. Ma, C. Tang, Y. Wang, L. Wei, Y. Z. Zhang, Z. X. Zhao, E. Baggio-Saitovitch, "Positron annihilation study on the stress-field pinning mechanism in (Eu,Y)-123 superconductors," Physica C, **314**, 55-68 (1999).

[23]Y. Shiohara, A. Endo, "Crystal growth of bulk high-T_c superconducting oxide materials," *Mater. Sci. Engineer.*, **R19**, 1-86 (1987).

[24]J. M. S. Skakle, "Crystal chemical substitutions and doping of $YBa_2Cu_3O_x$ and related superconductors," *Mater. Sci. Engineer.*, **R23**, 1-40 (1998).

[25]J. L. MacManus-Driscoll, "Materials Chemistry and Thermodynamics of $REBa_2Cu_3O_{7-x}$ " *Adv. Mater.*, **9**, 457-473 (1997).

[26]T. J. Haugan, J. C. Tolliver, J. M. Evans, J. W. Kell, "Crystal Chemical Substitutions of $YBa_2Cu_3O_{7-\delta}$ To Enhance Flux Pinning", in HTS Thin Film and More on Vortex Studies (Studies of High Temperatures Superconductors, Volume 49) (Ed. A. Narlikar), Nova Sci. Publishers, NY (2005).

[27]X.W. Cao, X.J. Xu, Z.H. Wang, J. Fang, R.L. Wang, H.C. Li, "Properties of upper critical field and pinning potential in epitaxial $GdBa_2Cu_3O_{7-\delta}$," *Physica C*, **282-287**, 1993-1994 (1997).

[28]L. Bejjit, A. Deville, M. Haddad, B. Gaillard, H. Noel, O. Monnereau, "Gadolinium Substitution for Yttrium and Chemical Preparation Effects in $YBa_2Cu_3O_7$: An EPR Study," *Ann. Chim. Sci. Mat.*, **23**, 217-220 (1998).

[29]Y. Feng, L. Zhou, J.G. Wen, N. Koshizuka, A. Sulpice, J.L. Tholence, J.C. Vallier, P. Monceau, "Fishtail effect, magnetic properties and critical current density of Gd-added PMP YBCO," *Physica C*, **297**, 75-84 (1998).

[30]G.F. Goya, "Magnetic Properties of the solid solution $(Y_{1-x}Gd_x)_2BaCuO_5$ $(0 \leq x \leq 1)$," *J. Mag. and Magn. Mat.*, **205**, 215-220 (1999).

[31]V.N. Narozhnyi, V.N. Kochetkov, "Influence of rare-earth ionic radius on the properties of Ni- and Fe-substituted R $Ba_2(Cu_{1-x}M_x)O_{7-y}$ systems (R=Y,Nd,Eu,Gd,Ho,Tm; *M*=Ni,Fe)," Phys. Rev. B, **53(9)**, 5856-5862 (1996).

[32]R.L.Wang, H.C. Li, B. Yin, J.W. Li, X.S. Ron, C. Dong, F. Wu, H. Chen, Z.X Zhao, "Critical current density and flux pinning in $Gd_{1-x}Y_xBa_2Cu_3O_{7-y}$ epitaxial thin films," *Physica C*, **250**, 55-58 (1995).

[33] H.H. Wen, Z.X. Zhao, R.L.Wang, H.C. Li, B. Yin, "Evidence for the lattice-mismatch-stress-field induced flux pinning in $(Gd_{1-x}Y_x)Ba_2Cu_3O_{7-\delta}$ thin films", *Physica C*, **262**, 81-88 (1996).

[34]Z.S. Peng, J.M. Hao, B. Yin, A.X. Zhao, Z.Q. Hua, B.C. Yang, "The Relationship of the Critical Current Density and the Lattice Deformation in $Y_{1-x}RE_xBa_2Cu_3O_{7-\delta}$ Epitaxial Films," *Physica C*, **282-287**, 2103-2104 (1997).

[35]A.R. Devi, V.S. Bai, P.V. Patanjli, R. Pinto, N.H. Kumar, S.K. Malik, "Enhanced critical current density due to flux pinning from lattice defects in pulsed laser ablated $Y_{1-x}Dy_xBa_2Cu_3O_{7-\delta}$ thin films", Supercond. Sci. Technol. **13**, 935-939 (2000).

[36]Y. Li, G. K. Perkins, A. D. Caplin, G. Cao, Q. Ma, L. Wei, Z. X. Zhao, "Study of the pinning behaviour in yttrium-doped Eu-123 superconductors," Supercond. Sci. Technol., **13**, 1029-1034 (2000).
[37]W. Wong-Ng, et. al., private communication and presented at DOE Peer Review, 2004.
[38] J. Shimoyama, Univ. of Tokyo, private communication, 2003.
[39]*Magnetic Susceptibility of Superconductors and Other Spin Systems*, Plenum Press, New York, 1991.
[40]Z. Chen, Y. Xue, Y. Su, J. Jin and Z. Tian, "Influence of Gd/Eu substitution on the local electron density and the superconductivity of $Y_{1-x}RE_xBa_2Cu_3O_{7-\delta}$ (R=Gd,Eu) systems," presented at CEC-ICMC, Keystone CO Sept. 2005.

Ni-20%Cr COATINGS ON BIAXIALLY TEXTURED COPPER AND COPPER–IRON ALLOY SUBSTRATES FOR $YBa_2Cu_3O_{7-x}$ COATED CONDUCTOR APPLICATIONS

Chakrapani V. Varanasi, Gerald R. Landis
University of Dayton Research Institute
Dayton, OH 45469

Paul N. Barnes, Jack L. Burke, Nicholas A. Yust, Timothy J. Haugan
Propulsion Directorate, Air Force Research Laboratory
Wright-Patterson AFB, OH 45433

ABSTRACT

To provide oxidation protection to Cu-based substrates during coated conductor processing, Ni-20%Cr coatings were investigated. These coatings were deposited on biaxially textured Cu and Cu-Fe alloy substrates by DC magnetron sputtering method at different processing conditions. The Ni-20wt% Cr coatings textured well when deposited at 500 °C and provided good oxidation protection to the underlying copper based substrates as determined by thermo-gravimetric analysis. Texture analyses and microstructures of the Ni-Cr coatings are presented.

INTRODUCTION

High temperature superconducting (HTS) $YBa_2Cu_3O_{7-x}$ (YBCO) coated conductors are presently being developed worldwide using metallic substrates on which buffer layers and YBCO coatings are applied [1-6]. Ni or Ni alloys are the most commonly used metallic substrates both in the textured substrate approach or rolling-assisted biaxially textured substrate (RABiTS) process [7], or the ion beam assisted deposition process (IBAD) [8]. However, copper or copper based alloys are an attractive, less costly alternative to Ni-based alloys as substrate materials for $YBa_2Cu_3O_{7-x}$ (YBCO) coated conductor applications. Cu has better lattice match (a=3.62 A°) with YBCO (a= 3.83 A°) than Ni (a= 3.52 A°), and Cu and Cu-alloy substrates with excellent biaxial texture better than Ni can be made [9-11].

Use of a thermally conductive copper substrate will have the benefit of more adequately conducting the heat away from a localized hot spot which might originate during the passage of high currents through the coated conductor while circumventing a normal zone in the HTS layer. If so, copper based substrates may act as a stabilization component of the HTS coated conductor by assisting in quench protection. If the YBCO layer is then electrically connected to the copper substrate via conducting buffer layers, the need for a copper current shunt can be reduced or even avoided. This can result in an increase in the engineering current density (J_e) of the coated conductor [12]. These advantages can make copper a superior substrate for the dc conductor as compared to Ni-based substrates.

Although some success has been achieved in demonstrating the use of Cu substrates, the oxidation of both copper and copper alloy substrates has been problematic in processing the YBCO coated conductors [9]. Deposition of oxidation protection layers on copper is an attractive approach to address this problem. However, the protective layer must not only provide oxidation protection, but also have a good lattice match, be able to grow with biaxial texture, remain effective (diffusion/alloying with copper during subsequent use), and preferably be non-magnetic. Ni overlayers have been deposited on Cu substrates by DC magnetron sputtering [13] and by electroplating [14] for subsequent growth of buffer layers and YBCO. However, Ni is magnetic and is also susceptible for oxidation during subsequent processing. Cantoni et al. [9] demonstrated that high J_c samples based on copper substrates can be made through proper selection of buffer layer architecture such as $LaMnO_3/MgO/TiN$ by inhibiting the oxidation of the substrates. There is a need for developing high rate of deposition of oxidation protection layers.

Ni-20wt% Cr has excellent oxidation resistance as compared to pure Ni and exists as a solid solution of Ni and Cr up to 20 wt% Cr. In this study, oxidation resistant, non magnetic Ni-20% Cr coatings have been deposited onto cube textured copper and copper–iron alloy substrates by DC magnetron sputtering to develop coatings with good texture and oxidation protection to the Cu and copper alloy substrates. In the literature, it has been demonstrated that YBCO coated conductors can be grown on Ni-7 at% Cr alloy substrates if proper steps are taken to avoid the formation of Cr_2O_3 [7, 15, 16]. It is expected that Ni-20wt%Cr coated Cu substrates also may require similar steps to be taken to process the coated conductors. As texture in Ni-20%Cr substrates is difficult to achieve, textured Ni-20% Cr coatings on textured Cu substrates offer an alternative. The work is presently ongoing to deposit the buffers and YBCO. As they are non magnetic, Ni-20% Cr targets facilitate the use of DC magnetron sputtering to achieve high rate of deposition. Texture development of the buffers and oxidation protection of the substrates were investigated for both Copper and Cu-Fe substrates at different deposition conditions and the results obtained in this study are presented.

EXPERIMENTAL

The Cu and Cu-Fe alloy substrates were processed by optimizing the thermo-mechanical treatments as described in detail elsewhere [10, 11]. The texture of the substrates and the coatings was determined by using a Philips 4 circle x-ray diffractometer. The full width half maxima (FWHM) of the phi and omega scans were determined from a Gaussian curve fitted to the data to eliminate the errors in the FWHM measurement. A field emission scanning electron microscope (FESEM)) was used to determine the grain structure of the substrates and the coatings. Prior to deposition, the substrates were cleaned in acetone and methanol. For the Ni-Cr coatings, a Denton DV-502A D.C. magnetron sputtering unit with a 3" Ni-Cr (80% Ni, 20% Cr) target was used with the rate of deposition being ~3 A°/min. The chamber was pumped down to a base vacuum of 3×10^{-6} torr. The substrate to target distance was about 6" and Ar was used as the sputtering gas with a pressure of 5 microns. The samples were heated by using a US thin film substrate heater model SU-200C and the temperature was determined by the use of a thermocouple. When the heater temperature was stabilized, the Ar gas was turned on and the sputtering process was initiated. Sputtering times for various durations were tried. Approximately 200-300 watts of power was applied. The sample thickness was measured using a Sycon STM100 quartz crystal rate monitor. Coatings up to 1 μm thickness were grown.

A thermal gravimetric analysis (TGA) instrument (TA instruments, SDT Q600), was used to compare the oxidation behavior of coated and uncoated substrates. For this test, the coated samples were prepared with Ni-Cr coatings on both sides of the Cu substrate. Both coated and uncoated samples were run in two separate identical runs. The temperature was ramped at 10 °C/min to 750°C in flowing oxygen gas at a 100 ml/min flow rate. The weight gain of the substrates was noted continuously during the heat treatment. As the substrates become oxidized, the weight will increase and this weight gain was used to compare the oxidation resistance.

RESULTS AND DISCUSSION

Ni- 20Wt% Cr COATINGS

Figure 1 shows the theta-two theta scan of a Cu substrate samples coated with Ni 20wt% Cr at 500 °C. As Ni and Cr form solid solutions up to 20 wt% Cr, the Ni (200) reflection at 51.8° was taken as indication of Ni-20 Wt% Cr coating texture. It was found that in the samples coated at 500 °C, c- axis texture of Ni –Cr can be obtained, whereas at the lower deposition temperatures such as 400 °C, and 300 °C reflections from other orientations were also observed. No extra peaks corresponding to Cr were observed. Figure 2 shows the x-ray phi scan of Cu and Ni-Cr coatings deposited at 500 °C. The full width half maximum (FWHM) of Ni-Cr coatings is 6.95° on a Cu substrate with a FWHM of 7.35 °, indicating that good in-plane alignment of the buffer layer is possible to obtain at this deposition temperature. Figure 3 shows the omega scans of the Ni-Cr coating with FWHM of 4.84 ° on a Cu substrate of FWHM 5.64°. In general, both the in-plane and out-of-plane texture of the Ni-Cr coatings were found to be relatively better than the starting Cu substrate. Figure 4 shows the microstructure of the coating at two different magnifications. The underlying grain structure of the copper can be seen through the buffer layer coating in Figure 4 a. The grains in the coating were found to be much smaller (200-300 nm) than the starting copper grains of copper which were 300-500 μm. Low temperature growth of 500 °C is believed to be the reason for the small grain structure observed in the coatings.

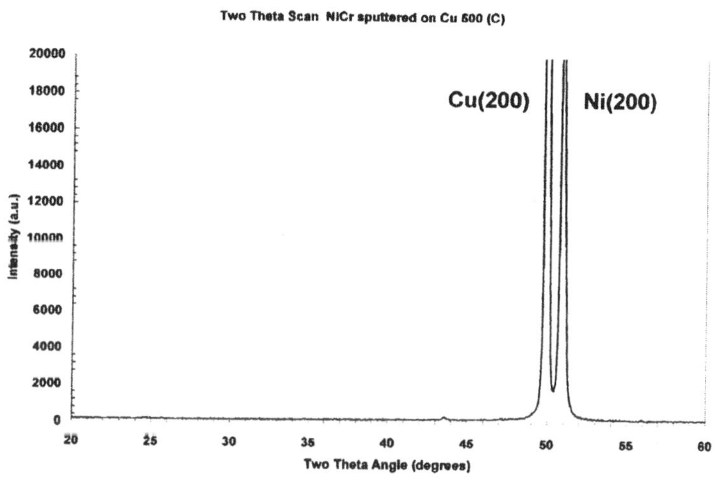

Figure 1. X-ray θ - 2 θ scans of Ni-20wt% Cr coatings deposited at 500 °C on textured Cu substrates.

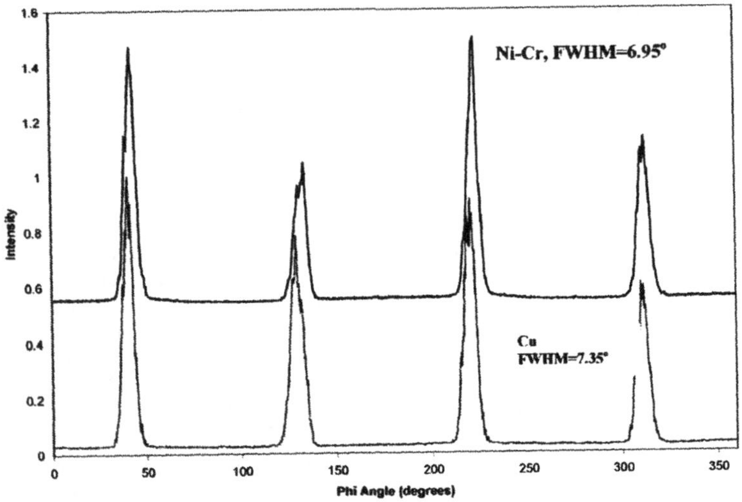

Figure 2. X-ray (111) Phi scan of Ni-20wt% Cr coating showing a FWHM = 6.95 ° on a Cu substrate with FWHM=7.35°

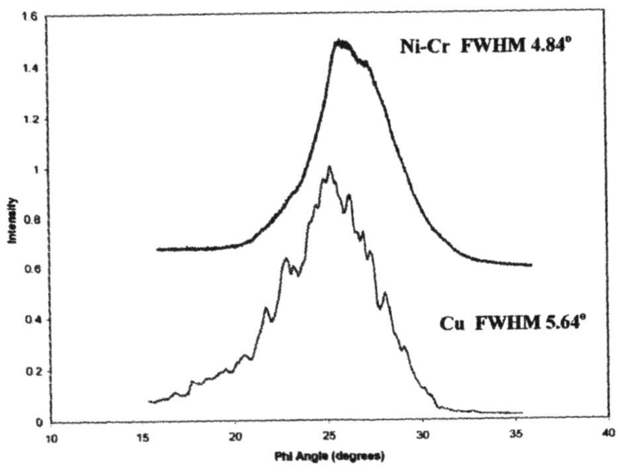

Figure 3. X-ray (100) Omega scan (parallel to rolling) of Ni-20wt% Cr coating FWHM = 4.84° and the Cu substrate FWHM=5.64°

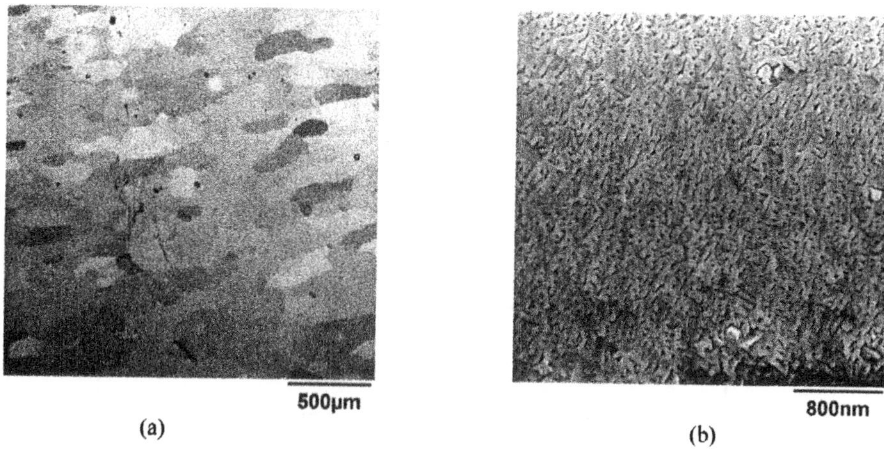

Figure 4. SEM picture showing a) grain structure of the copper substrate b) Grain structure of the Ni-20wt% Cr coating on this substrate

It has been reported that it is hard to get highly textured Ni alloy substrates, especially with high alloying concentrations, by using the rolling assisted biaxially textured substrate method due to the lowering of the stacking fault energy associated with the alloying [17]. For example, in the case of Ni – W alloys, Ni-5 wt% W is currently the highest amount of W that can be added to Ni and still be able to make highly textured alloy substrates. Although one would like to use the substrates of Ni-9 wt% W due to reduced magnetic nature [18], the high degree of biaxial texture has been to date, hard to achieve in these alloys due to the high amounts of W additions. The present result of obtaining highly textured Ni-Cr alloy coatings with 20 wt% of Cr indicates that the alloys which are difficult to texture by the standard RABiTS process, can probably be made by coating such alloys on textured Copper substrates and obtain the desired texture in alloy surfaces.

Figure 5 shows the TGA plot showing the percentage weight gain of both coated and uncoated substrates for the oxidation studies. It can be seen that the uncoated copper samples showed significant amounts of weight gain as compared to the Ni-Cr coated copper samples indicating the high degree of oxidation. In this experiment, the sides of the samples are not fully coated with Ni-Cr and some of the oxidation is attributed to this exposure as opposed to being completely in the Ni-Cr coating. This might be the reason for the observed slight weight gain in coated samples. At the end of the test the pure copper sample was completely oxidized and crumbled into pieces whereas the Ni-Cr coated piece stayed intact. This evidence suggests that the Ni-20wt% Cr coating deposited at the present deposition conditions can provide good oxidation protection.

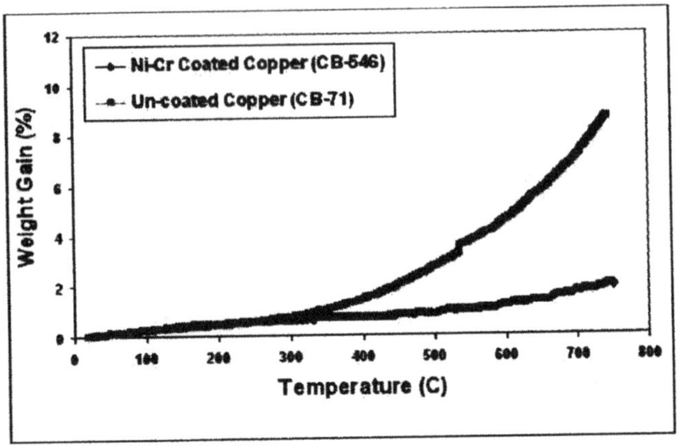

Figure 5. Percent weight gain vs. temperature plot of uncoated and Ni-20 wt% Cr coated Copper samples. (Data from the TGA experiment).

Ni-Cr coatings on Cu-Fe samples processed at 500 °C also showed similar results as observed with pure copper substrates. Figure 6 shows the phi scan of the Ni-Cr coatings on Cu-Fe samples showing that FWHM is 6.46° on a substrate that had FWHM as 5.84°. Figure 7 shows the omega scan showing an out-of-plane FWHM of 9.8° while the substrate had a FWHM of 5.96° on the substrate. It appears that the texture of the Ni-Cr seems to be poorer than the Cu-Fe substrates as opposed to the pure Cu substrates. Although the reasons are not known at present, it is thought that the surface roughness played an important role in determining the quality of the textured coatings. The copper substrates used in this study were much smother than Cu-Fe substrates and this is quite possibly responsible for the better textured Ni-Cr coatings on Cu substrates as compared to Cu-Fe substrates.

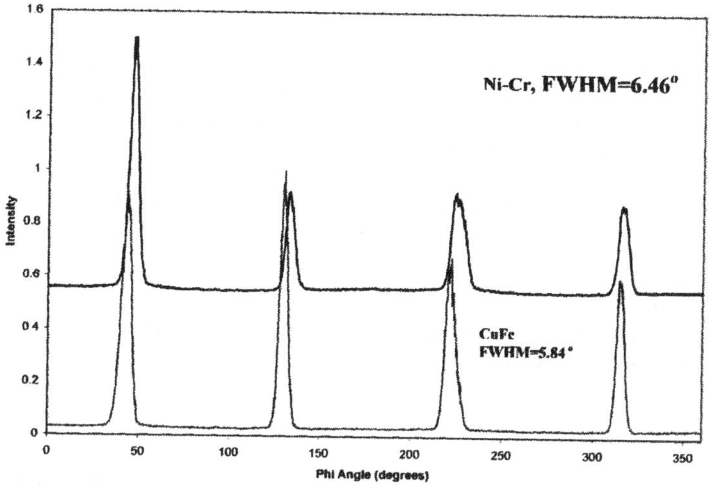

Figure 6. X-ray (111) Phi scans of Ni-20wt% Cr coating showing a FWHM= 6.46° on a Cu-Fe samples with FWHM=5.84°.

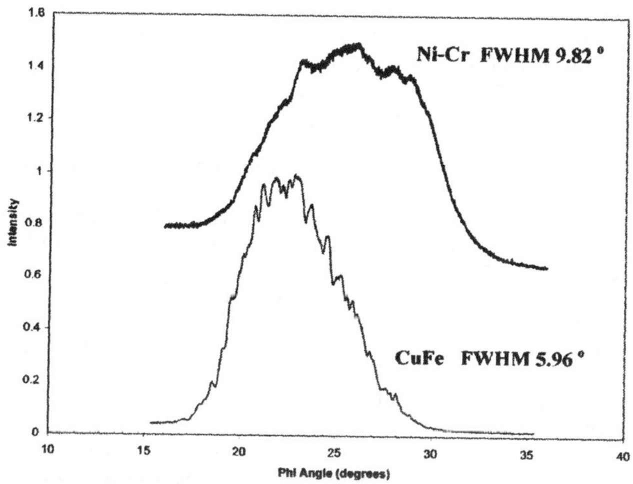

Figure 7. X-ray (100) Omega scan (parallel to rolling) of a Ni-20wt% Cr coating FWHM = 9.82° and a Cu-Fe substrate FWHM=5.96°.

CONCLUSIONS

Biaxially textured Cu and Cu-Fe substrates were processed by optimizing the thermo-mechanical processing schedules. Biaxially textured Ni-20 wt % Cr non-magnetic buffer layers were grown successfully on both biaxially textured Cu and Cu-Fe substrates using dc magnetron sputtering by optimizing the deposition conditions. Ni-20 wt% Cr coatings provided good oxidation protection for the copper substrates as determined by TGA analysis in oxygen. These layers can be used as an intermediate layers for the subsequent growth of the coated conductors.

ACKNOWLEDGEMENTS

The authors appreciate the assistance provided by Lyle Brunke and John Murphy with the annealing treatments of the substrates.

REFERENCES

[1]Verebelyi, U. Schoop, C. Thieme, X. Li, W. Zhang, T. Kodenkandath, A.P. Malozemoff, N. Nguyen, E. Siegal, D. Buczek, J. Lynch, J. Scudiere, M. Rupich, A. Goyal, E.D. Specht, P. Martin, and M. Paranthaman : Supercond. Sci. Technol., **16**, L19 (2003).
[2]Selvamanickam V, Lee HG, Li Y, Xiong X, Qiao Y, Reeves J, Xie J, Knoll A, Lenseth K, Physica C, **392**, 859(2003).
[3]Kakimoto K, Iijima Y, Saitoh T, Physica C, **392**,783 (2003).
[4]Usoskin A, Freyhardt HC, Issaev A, Dzick J, Knoke J, Oomen MP, Leghissa M, Neumueller HW, ,IEEE Trans on Appl. Supercond. **13**, 2452 (2003).
[5]Selvamanickam V, Knoll A, Xie Y, Li Y, Chen Y, Reeves J, Xiong X, Qiao Y, Salagaj T, Lenseth K, Hazelton D, Reis C, Weber C, Yumura H, Applied Superconductivity Conference, October 4-8 (2004).
[6]Iijima Y, Kakimoto K, Sutoh Y, Ajimura S, Saitoh T, ,Applied Superconductivity Conference, October 4-8 (2004).
[7]Goyal A, Lee DF, List FA, Specht ED, Feenstra R, Paranthaman M, Cui X, Lu SW, Martin PM, Kroeger DM, Christen DK, Kang BW, Norton DP, Park C, Verebelyi DT, Thompson JR, Williams RK, Aytug T, Cantoni C, , Physica C **357-360**, 903 (2001).
[8]S.R. Foltyn, P.N. Arendt, R.F. DePaula, P.C. Dowden, J.Y. Coulter, J.R. Groves, L.N. Haussamen, L.P. Winston, Q.X. Jia, M.P. Maley, Physica C **341-348,** 230 (2000).
[9]Cantoni, D. K. Christen, E.D. Specht, M. Varela, J.R. Thompson, A. Goyal, C. Thieme, Y. Xu, S. J. Pennycook, Supercond. Sci. Technol.**17,** S341-S344 (2004).
[10]Nicholas A Yust, Rama Nekkanti, Lyle B. Brunke, Raghavan Srinivasan, Paul N Barnes, Supercond. Sci. Technol.**18**, 9 (2005)
[11]Chakrapani V. Varanasi, Paul N. Barnes, Nicholas A. Yust, , submitted for publication (2005)
[12]P.N. Barnes, G.L. Rhoads, J.C. Tolliver, M.D. Sumption, and K.W. Schmaeman: IEEE Trans. Magn. **41**, 268 (2005).
[13]N.A. Rutter, A.Goyal, C.E. Vallet,F.A. List, D.F. Lee, L. Heatherly, D.M. Kroeger, Supercond. Sci. Technol.**17,** 527 (2004).

[14]Y.X. Zhou, L. Sun, X. Chen, H. Fang, P.T. Putman, K. Salama, Supercond. Sci. Technol.**18,**107 (2005).
[15]J.R. Thomson, A.Goyal, D.K. Christen, D.M. Kroeger, Physica C **370**,169 (2002)
[16]R.I. Tomov, A. Kursumovic, M. Majoros, B.A. Glowacki, J.E. Evetts, A. Tuissi, E. Villa, M. Zamboni, Y.Sun, S. Tonies, H.W. Weber, Physica C **383**,323 (2003)
[17]N. Brown, Trans, AIME **215,** 530 (1961).
[18]A.O. Ijaduola, J.R. Thomson, A.Goyal, C.L.H. Thieme, K. Marken, Physica C, **403**,163 (2004).

COMPARISON OF YBCO THIN FILMS FABRICATED USING DIFFERENT APPROACHES

S.M. Mukhopadhyay, J.H. Su, V. Chintamaneni, S. Vemulakonda, and P.P. Joshi
Mechanical and Materials Engineering, Wright State University
Dayton, OH 45435

ABSTRACT

Three popular fabrication approaches are currently able to produce high quality $YBa_2Cu_3O_{7-\delta}$ (YBCO) films, including pulsed laser deposition (PLD), metal-oxide chemical vapor deposition (MOCVD), and metal-organic deposition using trifluoroacetate (TFA-MOD). A comparison of chemical and compositional profiles of films grown by these methods has been performed using x-ray photoelectron spectroscopy (XPS) and scanning electron microscopy (SEM) in conjunction with chemical etching to obtain composition and microstructure information at different depths into the film. The durability and stability of different films have also been compared by measuring etch rates. It is seen that samples grown by the TFA-MOD method have significantly higher etch rates than others, even though they have comparable J_c values. Fine porosity seen in MOD films is sometimes suspected of helping in flux pinning, but is bound to reduce the mechanical robustness. TFA-MOD is the only non-vacuum approach, which would provide a cost advantage during scale up of these films. However, the process is very slow, and notoriously sensitive to heating rates. With this in mind, fundamental study on TFA-MOD film evolution was performed to correlate heating rates with material property. In parallel an alternate non-vacuum MOD processing route is being investigated with the idea of creating films denser and faster than the current TFA-MOD process. Films prepared by this new approach have been compared with those grown by the TFA-MOD process, and indicate future promise.

INTRODUCTION

There are many processes for making $YBa_2Cu_3O_{7-\delta}$ (YBCO) superconducting films. Typical ones are pulsed laser deposition (PLD), metal-organic chemical vapor deposition (MOCVD), and metal-organic deposition (MOD). PLD is a vacuum process that is most reliable and widely used to achieve J_c values higher than 1 MA/cm^2 [1-3]. The principle concern about this method is the cost of scaling up. A lower vacuum and therefore potentially lower cost technique is MOCVD, which has been developed, particularly by Fujikura[4] and IGC-SuperPower [5]. While the deposition rate is rapid, uniformity over large areas and the precursor cost, involving complex organic molecules, are still significant challenges. The most successful, non-vacuum method is so-called metal-organic deposition technique using trifluoroacetate solution (TFA-MOD), first initiated by Gupta *et al.*[6] in 1988 and optimized by McIntyre *et al.*[7] This is an efficient method of producing large-scale, biaxially-textured superconducting films at lower cost than physical and chemical vapor deposition techniques.

The main draw backs for the TFA-MOD process are: (i) long process time, (ii) porosity in the final film, and (iii) evolution of highly corrosive HF during process. In the TFA-MOD process, a very slow calcination (pyrolysis process, usually > 20 h) is needed to control the decomposition of fluoroacetates. Another issue with the TFA-MOD process is the evolution of highly corrosive HF, which is a by-product during high temperature conversion process and results in a porous microstructure for TFA-MOD processed YBCO films.

From the viewpoint of the scale-up applications, alternate approaches with short calcination and lower porosity will be useful to industry. Starting with nanoparticles of the three cations, Y, Ba, and Cu, may be an option. This process will decrease the porosity of the final films because the gaseous by-products could be significantly reduced and may be completely eliminated. It is also expected to significantly shorten heat treatment time and lower the processing temperatures.

Microstructures, grain sizes, and electric properties of YBCO films significantly depend on different processing techniques. In this paper, we firstly investigate the influence of various processing techniques on the microstructures and peeling rates of YBCO films. We then report the influence of heating rate during calcining in the TFA-MOD process on the morphology and the electric transport properties of YBCO films. A new route for the production of YBCO thin film, which uses a colloidal suspension of Y-Ba-Cu-O nanoparticles as precursor instead of metal trifluoroacetates has also been reported.

EXPERIMENTAL DETAILS

Surface chemistry was characterized by x-ray photoelectron spectroscopy (XPS) in a Kratos-Axis Ultra System using monochromatized Al Kα photons. XPS data not only identifies what elements are present at the top 1-5 nm of the surface, but also distinguishes chemical bonding states of the identical atoms. Microstructures were studied by a JOEL 35 scanning electron microscope with 8 nm resolution. Phase compositions were analyzed via Cu-Kα powder x-ray diffractometry (XRD). Samples prepared by PLD, MOCVD and MOD, which were obtained from commercial companies and national laboratories, were used to investigate the influence of fabrication approaches on microstructure and chemistry through the entire thickness of films. The detailed information of those samples used in this study is shown in Table 1. To study depth profile of microstructure and chemistry of the films, a layer-by-layer etching technique was used. Bromine ethanol was selected to etch YBCO films because bromine does not affect the microstructure and chemical profile[8]. After each etching cycle, the fresh surface exposed was characterized using XPS and SEM.

Samples used to study the chemistry evolution and the influence of the heating rate in the TFA-MOD process, were prepared in house. Fig. 1 shows the synthesis process for the TFA-MOD approach. A trifluoroacetate solution precursor was prepared by dissolving the acetates of Y, Ba, and Cu in a 1:2:3.05 cation ratio in de-ionized water with 10% excess TFA acid at room temperature. The resulting solution was dried overnight in air to yield a glassy blue residue. The residue was dissolved in sufficient methanol to give a solution with a 1.0 M cation concentration. The TFA precursor solution was then spin coated onto LAO single crystals at 3600 rpm for 2 min. Spin-coated gel films were calcined at the temperature up to 400 °C, as seen in Fig. 2(a). All calcined films were subsequently converted to YBCO by annealing in humid argon/oxygen gas mixtures at 800 °C for 1 h (Fig. 2(b)) in a low oxygen partial pressure ($P_{(O2)}$ = 200 ppm). For oxygenation, the films were held at 450 °C for 90 min followed by furnace cooling to room temperature in flowing oxygen.

For study of the nano-fabrication process, the precursor used is a colloidal suspension of Y-Ba-Cu-O nanoparticles in an organic solvent. This precursor solution was provided by an industrial laboratory. Similar to the TFA process, the precursor films were spin-coated onto LAO

Table 1. Description of samples for bromine etching experiments.

Sample ID	Top Layer	Buffer layer	Substrate	Deposition technique
1	YBCO	CeO_2/YSZ	Inconel	PLD
2	YBCO	CeO_2/YSZ	Haste Alloy	MOCVD
3	YBCO	CeO_2/YSZ/ CeO_2	Ni-W	MOD

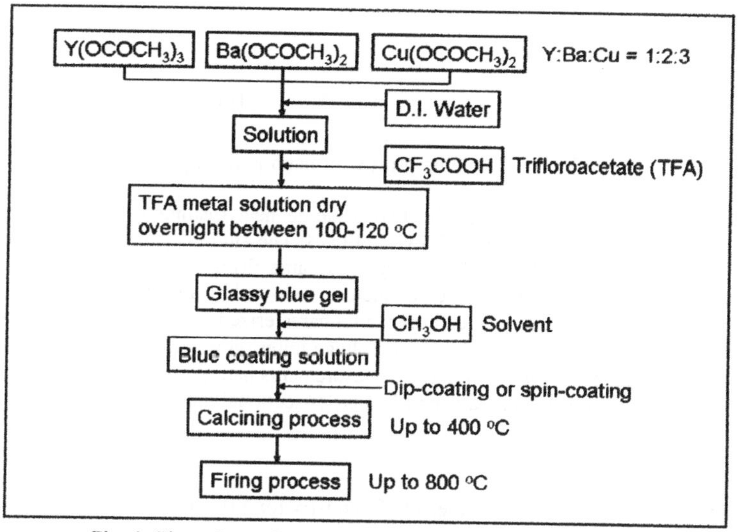

Fig. 1. Flow chart of TFA-MOD process.

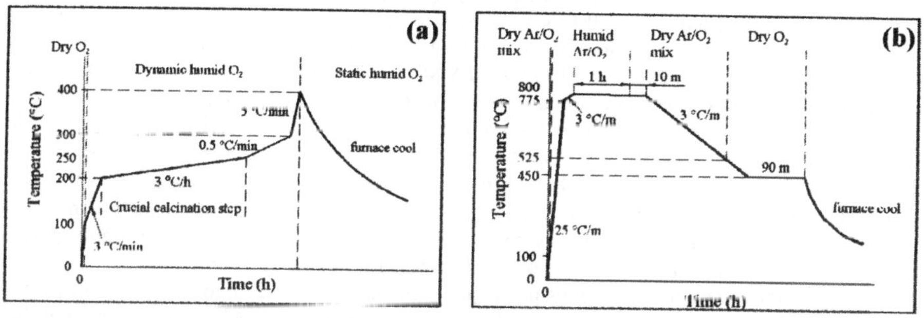

Fig. 2. Heating profiles for TFA-MOD process: (a) calcination and (b) high temperature annealing.

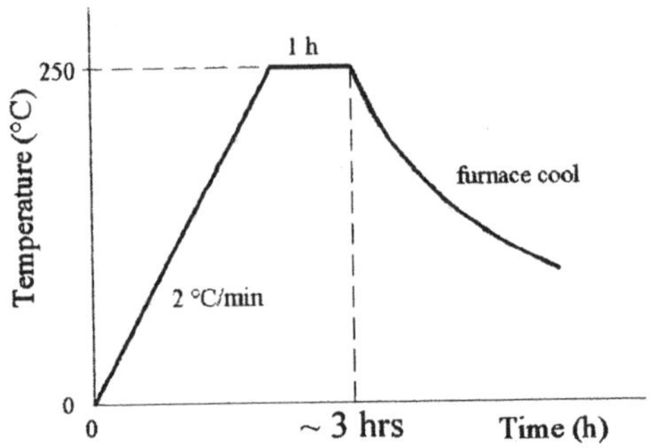

Fig. 3. Calcining profile for nano-particle process.

substrates and then were heat treated by two separate heat treatment steps. The calcination profile for the nano-particle process is shown in Fig. 3. The entire calcination time is ~ 3 hours, compared to ~ 20 hours for TFA-MOD process. A same firing profile as that for TFA-MOD process is used to crystallize YBCO superconductors, followed by an oxygen annealing.

The electrical properties of the films were characterized by measuring T_c and I_c using a standard four-point configuration. A 1 μV/cm voltage criterion was used to determine I_c at 77 K and self field. The samples prepared for T_c/I_c measurements had typical dimension of 10 mm × 3 mm. Film thicknesses were estimated from SEM cross-sectional images.

RESULTS AND DISCUSSION

Dependence of Microstructure, Etch rate, and Chemistry for YBCO Films on Fabrication Processes

Fig. 4 shows the microstructures of YBCO films synthesized by three methods, PLD, MOCVD, and TFA-MOD. The PLD sample showed a dense microstructure, but with micro-cracks. Micro-cracks are not present at the MOCVD sample, but many needle-like grains appeared, indicating a formation of a/b-oriented YBCO grains. The film deposited by TFA-MOD is porous with micro-cracks. The porosity is due to the escape of HF during the calcining and firing processes. The peeling rates of all three films were estimated by dividing the film thickness by total etch time. SEM and XPS were used to confirm the removal of YBCO layer. It was found that the TFA-MOD sample had the fastest peeling rate of ~ 0.006 μm/s, while PLD and MOCVD samples had an almost same rate of ~ 0.002 μm/s; three times lower than that of the MOD sample. A faster peeling rate for MOD samples maybe is due to the highly porous nature of MOD coatings and predicts lack of robustness for long length cables.

Fig. 5 shows the concentrations of surface elements vs. etch time for three types of films. The increase of the Ce peak, together with the decrease of Y, Ba, Cu peak, indicates the interface of YBCO/CeO_2. It is clearly that the MOD film has the shortest etch time of ~ 150 seconds, while the PLD and MOCVD films have the similar ones of ~ 500 seconds. Note that all three types of films have almost same film thickness, about 1.5 μm.

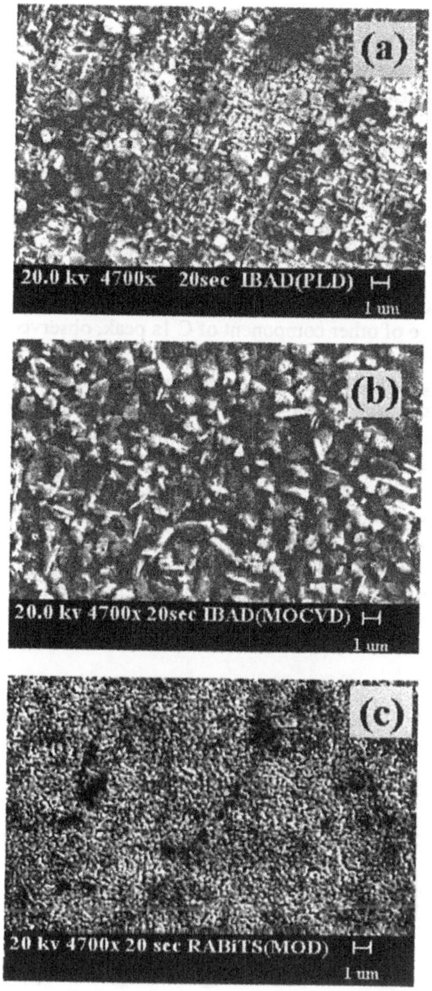

Fig. 4. SEM images of YBCO films synthesized by (a) PLD, (b) MOCVD, (c) TFA-MOD processes.

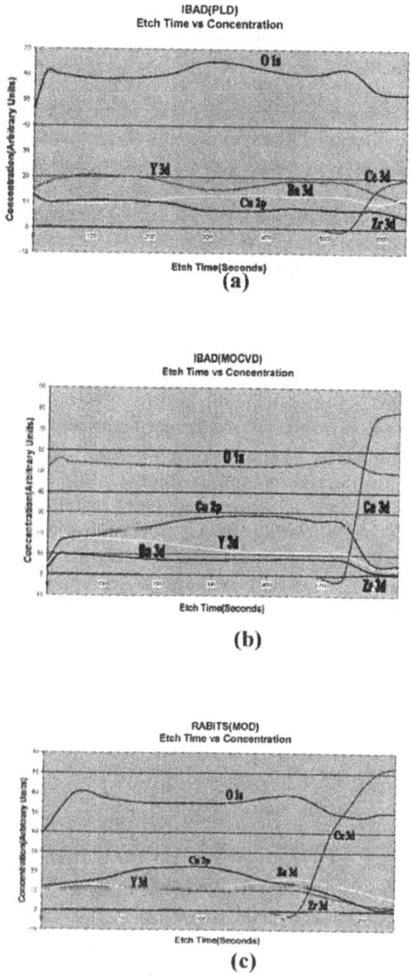

Fig. 5. Concentration of surface elements vs. etch time for YBCO films synthesized by (a) PLD, (b) MOCVD, (c) TFA-MOD processes. Note that the MOD processed sample was etched much faster compared to PLD and MOCVD samples.

Microstructure and Chemistry for TFA-MOD Process

TFA-MOD process includes two separate heat treatment steps, calcining and firing processes, as shown in Fig. 2. A very slow pyrolysis process (usually > 20 h) is needed to control the decomposition of metal trifluoroacetates at 200-250 °C during calcining. This is typical of organic chemical reactions that break a covalent bond. During the calcining process, metal trifluoroacetates decompose entirely and carbon content is expelled to give precursor film. XPS data are used to identify exactly where and how the metal fluorides changes to oxyfluorides and metal oxide. Fig. 6 shows the comparison of C 1s XPS spectra obtained on the films before and after the calcining process. Before calcination, the C 1s has several components. The component occurring at 291 eV is due to the $-CF_3$ group, where the one at 287.5 eV is –COOH. Note that after calcining process these two components become very weak, indicating decomposition of fluoroacetates group during calcination. The presence of other component of C 1s peak, observed on the films both before and after calcination, can be attributed to C-O-R and/or the atmospheric contaminants absorbed during sample transfer. Fig. 7 shows the F 1s spectra obtained on the films before and after calcination at 10 °C/h. The stronger component of fluorine (686.5 eV) in the film before calcination is attributed to $-CF_3$ group that is very strongly diminished after the calcining process. After calcination, most of fluoride has been found bonded with barium (BaF_2 compound), as reported by Gupta *et al.*[6]

Fluoride-containing calcined film was fired and annealed during the firing process to crystallize YBCO superconductors. Significant porosity is seen in the final films due to the escape of HF during process. To control the porosity, the heating rate is known to be a crucial parameter and an important variable for optimization. The following discuss is focus on the influence of heating rate during calcining process (200-250 °C) on the morphology and the electric transport properties of TFA-MOD processed films.

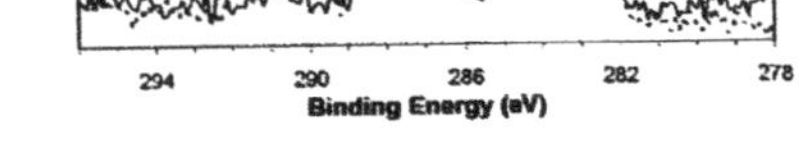

Fig. 6. C 1s XPS spectra obtained for a calcined film in the TFA-MOD process.

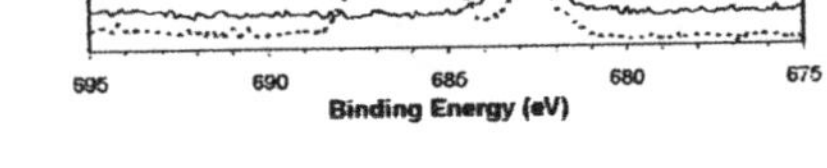

Fig. 7. F 1s XPS spectra obtained for a calcined film in the TFA-MOD process.

Influence of Heating Rate in TFA-MOD Process

To investigate the influence of heating rate, three different heating rates of 1.5, 3, and 10 °C/h were selected between 200-250 °C. Fig. 8 shows the surface morphologies of the calcined films heat treated at (a) 1.5 °C/h, (b) 3 °C/h, and (c) 10 °C/h. The film calcined at 3 °C/h shows a smooth and uniform surface (Fig. 8(b)), while a granular structure with grain size ~ 800 nm in diameter was observed on the film calcined at 1.5 °C/h (Fig. 8(a)) indicating phase separation in nano-scale. Araki *et al.*[9] has reported that almost all Cu in the precursor films convert to CuO nanocrystallites during calcination. For slowly heated films, an average grain size of 23.7 nm can be observed. It is hereby believed that the grains showed in Fig. 8(a) are the clusters of CuO nanocrystallites that had sufficient time to separate as a second phase. The film calcined at 10 °C/h showed an uneven surface with porous features (Fig. 8(c)), indicating many unclosed venting paths due to rapid escape of HF. Fig. 8(c) evidenced that the fastest heating rate during calcination leads to rapid HF generation, thus forming a porous feature which spoiled the uniformity of calcined films.

Surface morphologies of final films are shown in Fig. 9. A smooth surface characterized with the *c*-axis oriented microstructure was observed in the 3 °C/h heated films, as shown in Fig. 9(b). The slower heating rate of 1.5 °C/h, however, results in a large density of *a*-axis oriented grains, evidenced by needle-shaped contents seen in Fig. 9(a). Note that significant *a*-axis oriented grains appearing in the slow heated YBCO films have been also reported by Teranishi *et al.*[10] and Siegal *et al.*[11]. No *a*-axis oriented grains were observable on the samples heated at faster rates. Compared to that of 3 °C/h, the faster heating rate of 10 °C/h, however, results in bigger pores and a higher pore density (Fig.9(c)), as expected from the microstructure observation in Fig. 8(c).

Table 2 summaries the influence of heating rate on surface morphologies and electrical properties for the films calcined at various rates at 200-250 °C. Samples calcined at 3 °C/h and 10 °C/h exhibited almost the same $T_c(R{=}0)$ values (88.5 K and 88.4 K, respectively) and sharp transition width ΔT_c (1.2 K and 1.4 K, respectively), whereas the sample calcined at 1.5 °C/h had a lower $T_c(R{=}0)$ value of 87.3 K and a relatively broader ΔT_c (2.7 K). The transport properties at 77 K, self field for all the samples are also shown in Table 2. The 3 °C/h heated sample had the highest J_c value of 1.3×10^6 A/cm^2. The fact of lowest normal-state resistivity at 100 K and highest J_c value for the 3 °C/h heated samples correlates well with texture and microstructural investigations. It is clear that the surface morphology and electrical properties of as-prepared films depends on the heating rate.

Table 2. Summary of surface morphology and electrical properties vs. heating rates.

Heating rate	Surface morphology		Electrical properties			
	Precursor film	Final reacted film	$T_c(0)$ (K)	ρ(100 K) (μΩ·cm)	ΔT_c (K)	J_c (MA/cm^2)
1.5 °C/h	Granular structure with grain size ~ 800 nm	Large density of *a*-axis grains. Average film thickness: 460 nm.	87.3	265	2.7	0.2
3 °C/h	Smooth and uniform surface	No *a*-axis grains observable. Average film thickness: 450 nm.	88.5	130	1.2	1.3
10 °C/h	Uneven surface decorated with porous feature	No *a*-axis grains observable. Higher density of pores and bigger pores. Average film thickness: 420 nm.	88.4	184	1.4	0.6

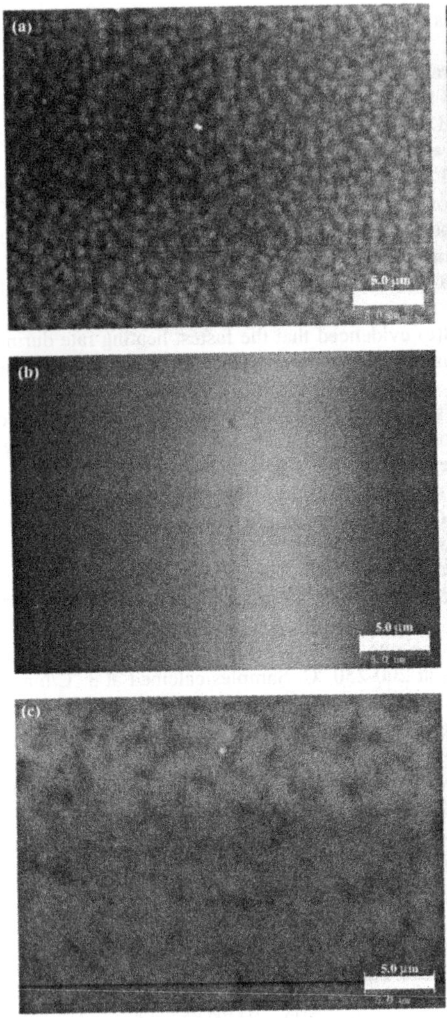

Fig. 8. SEM micrographs of the calcined films heated at (a) 1.5 °C/h, (b) 3 °C, and (c) 10 °C/h.

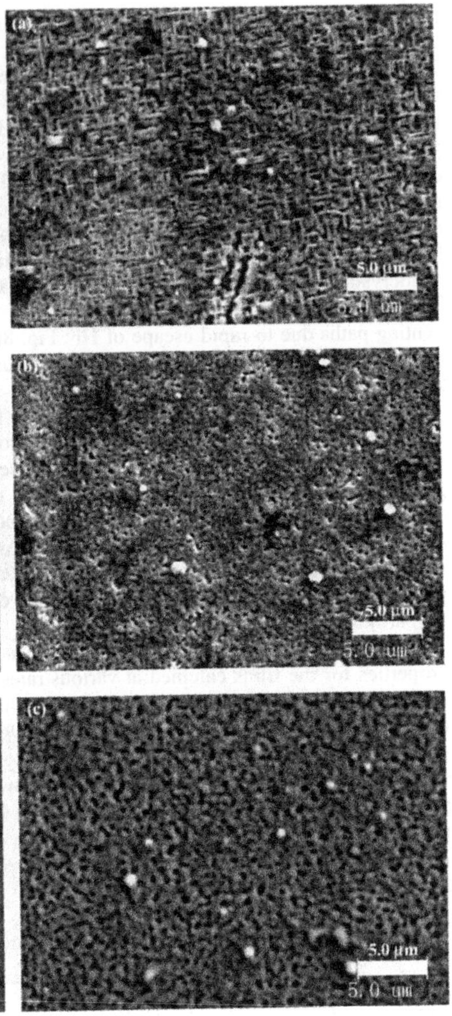

Fig. 9. SEM micrographs of the YBCO films calcined at (a) 1.5 °C/h and (c) 10 °C/h.

Fabrication of YBCO Films using Nano-Particles as Precursor

Fig. 10 shows the comparison of the Y 3d, Ba 3d, Cu 2p, and F 1s XPS spectra obtained from the films calcined by TFA-MOD process and nano-particle process. It is clearly that Y and Ba show difference in bonding states. In the TFA-MOD process, Y and Ba are still fluorides or oxyfluorides, while in the nano-particle process, Y and Ba are oxides. The peak shapes of Cu 2p look very similar, indicating that Cu is bonded with O in the both calcining processes. After calcining, the TFA-MOD processed films still had a significant F existing, while for the nano-particle process, only trivial F exists in the samples.

After high temperature annealing, the as-prepared films exhibited high phase purity YBCO with noticeable c-axis texture, as seen in Fig. 11. The reacted film show a much denser microstructure (Fig. 12), indicating much lower porosity than TFA-MOD processed films (Fig. 9). Fig. 13 displays the temperature dependence of the electrical resistivity showing a T_c (R = 0) ~ 82 K, which is about 6 K lower than the typical T_c values of TFA-YBCO films. I_c data are still not achieved, which may be due to the lower T_c value. Optimization of synthesis parameters is expected in the future work.

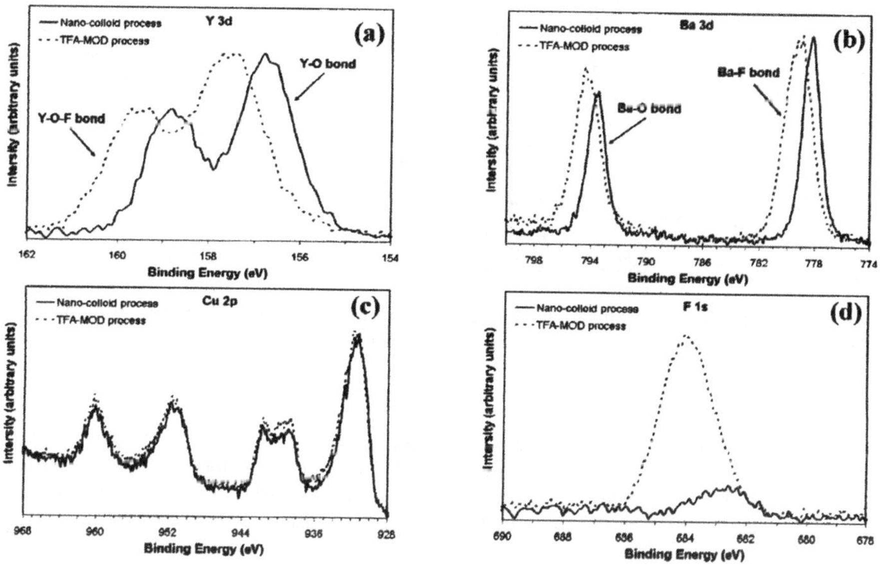

Fig. 10. Comparison of the (a) Y 3d, (b) Ba 3d, (c) Cu 2p, and (d) F 1s XPS spectra obtained from the films calcined by TFA-MOD process and nano- particle process.

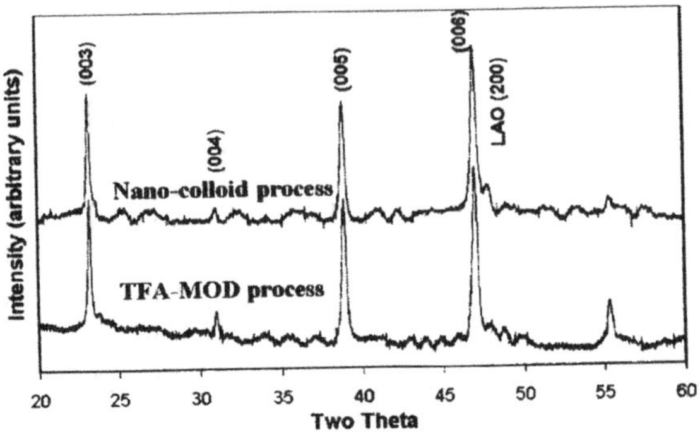

Fig. 11. XRD patterns for YBCO films synthesized by TFA-MOD and nao-particle processes.

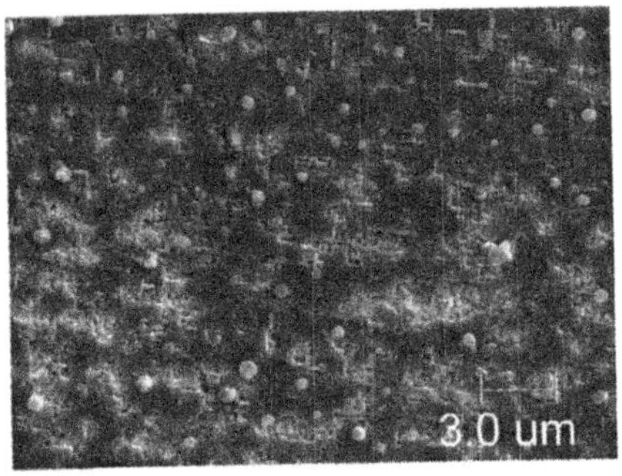

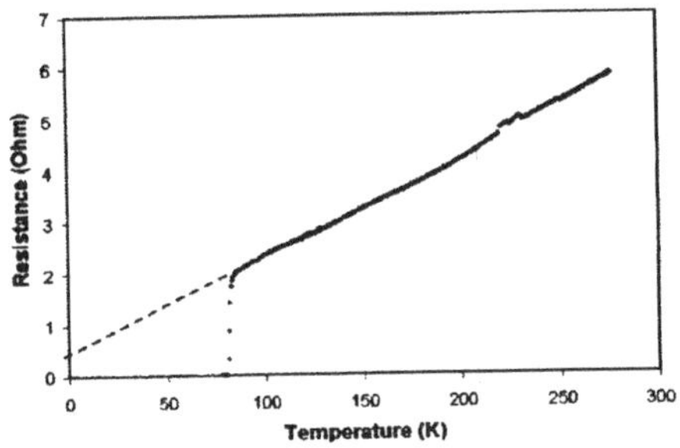

Fig. 12. SEM image of a final film using nano-particle process.

Fig. 13. T_c data for a final film using nano-particle process.

CONCLUDING REMARKS

The influence of three different synthesis methods, i.e., PLD, MOCVD, and TFA-MOD, on microstructure, chemistry, and etch rate of YBCO coated conductors has been investigated. PLD samples showed a dense microstructure, while the TFA-MOD processed film is porous, which is in agreement with the etch rate measurements; a faster peeling rate for MOD samples was observed.

Porosity is a nature of the TFA-MOD processed films due to the escape of HF during process, which degrades J_c values. Heating rate during calcination is known to be a crucial parameter to control the microstructure. A slower heating rate results in phase separation in the calcined films and causes a large density of *a*-axis oriented grains in the final reacted films. A faster heating rate results in a higher pore density together with a bigger average pore size.

Nano-particle process is promising because of its expected shorter processing time with lower porosity. We have successfully synthesized the c-axis-oriented YBCO films using this process. As-prepared films show a denser microstructure with a much shorter calcination time, compared to those prepared by TFA-MOD process. Our work shows that the fabrication approach using nano-scaled particles is attractive for economical manufacturing of long-length, second-generation YBCO coated conductors. Future work needs to be done on the optimization of processing parameters.

REFERENCES

[1]T.G. Holesinger, S.R. Foltyn, P.N. Arendt, H. Kung, Q.X. Jia, R.M. Dickerson, P.C. Dowden, R.F. DePaula, and J.Y. Coulter, "The microstructure of continuously processed $YBa_2Cu_3O_y$ coated conductors with underlying CeO_2 and ion-beam-assisted deposition/yttria-stabilized zirconia buffer layers," *J. Mater. Res.*, **15**, 1110 (2000).

[2]Y. Iijima and K. Matsumoto, "High-temperature-superconductor coated conductors: technical progress in Japan," *Supercond. Sci. Technol.*, **13**, 68 (2000).

[3]A. Usoskin, J. Knoke, F. Garcia-Moreno, A. Issaev, J. Dzick, S. Sievers, and H.C. Freyhardt, "Large-area HTS-coated stainless steel tapes with high critical currents," *IEEE Trans. Appl. Supercond.*, **11**, 3385 (2001).

[4]K. Onabe, S. Nagaya, T. Shimonosono, Y. Iijima, N. Sadakata, T. Saito, and O. Kohno, "Fabrication of YBCO superconducting tapes by continuous MOCVD technique," *Proc. 16th Int. Cryogenic Engineering Conf./Int. Cryogenic Materials Conf.*, ed. T. Haruyama, T. Mitsui, and K. Yamafuji, Oxford, UK: Elsevier Science, 1997, p. 1413.

[5]V. Selvamanickam, G. Carota, M. Funk, N. Vo, P. Haldar, U. Balachandran, M. Chudzik, P. Arendt, J.R. Groves, R. DePaula, and B. Newnam, "High-current Y-Ba-Cu-O coated conductor using metal organic chemical-vapor deposition and ion-beam-assisted deposition," *IEEE Trans. Appl. Supercond.*, **11**, 3379 (2001).

[6]A. Gupta, R. Jagannathan, E.I. Cooper, E.A. Giess, J.I. Landman, and B.W. Hussey, "Superconducting oxide films with high transition temperature prepared from metal trifluoroacetate precursors," *Appl. Phys. Lett.*, **52**, 2077 (1988).

[7]P.C. McIntyre, M.J. Cima, and M.F. Ng, "Metalorganic deposition of high-J_c $Ba_2YCu_3O_{7-x}$ thin films from trifluoroacetate precursors onto (100) $SrTiO_3$," *J. Appl. Phys.*, **68**, 4183 (1990).

[8]S.M. Mukhopadhyay and C. Wei, "Interface modification by Br in the Processing of Superconducting Composites," *Ceramic Transactions*, **102**, 217 (1998).

[9]T. Araki and I. Hirabayashi, "Review of a chemical approach to $YBa_2Cu_3O_{7-x}$ coated superconductors-metalorganic deposition using trifluoroacetates," *Supercond. Sci. Technol.*, **16**, R71 (2003).

[10]R. Teranishi, T. Honjo, Y. Tokunaga, H. Fuji, J. Matsuda, T. Izumi, A. Yajima, and Y. Shiohara, "Fabrication of YBCO film by TFA-MOD process at low-pressure atmosphere ," *Phys. C*, **412-414**, 920 (2004).

[11]M.P. Siegal, J.T. Dawley, P.G. Clem, and D.L. Overmyer, "Improving chemical solution deposited $YBa_2Cu_3O_{7-\delta}$ film properties via high heating rates," *Phys. C*, **399**, 143 (2003).

Bio-Interfaces

BIOMECHANICAL FUNCTION OF THE DENTINO-ENAMEL JUNCTION REFLECTS GENETIC CONTROL

S. N. White, V. G. Miklus, A.Y. Ngan, and K. Potter
UCLA School of Dentistry
CHS 23-010
Los Angeles, CA 90095-1668

W. Luo, M. L. Snead, and M. L. Paine
Center for Craniofacial Molecular Biology
University of Southern California School of Dentistry
2250 Alcazar Street CSA 103
Los Angeles, CA 90033

ABSTRACT

The dentino-enamel junction (DEJ) unites stiff, brittle enamel to tough, flexible dentin. Early near-DEJ dentin and enamel is known to differ from bulk tissues. We propose that transgenic animal models can be used to investigate the roles of specific genes in the unique mechanical properties of the DEJ.

Objectives: To evaluate the biomechanical function of the DSPP gene, expressed by odontoblasts and early ameloblasts, that expresses DSP and DPP proteins, on the DEJ by comparing the fracture mechanics of over-expressing transgenic mice to wild type mice; and to determine if Type I collagen, $[\alpha1(I)]2[\alpha2(I)]$, the dominant dentin protein, is responsible for toughening the DEJ by comparison of wild type mice with the transgenic Col1α2oim homotrimer mouse $[\alpha1(I)]3]$.

Methods: Tteeth of wild type and transgenic DSP, DPP, Col1α2oim mice were embedded, polished, and micro-indented. Bulk enamel hardness, bulk enamel fracture toughness, bulk dentin hardness, and DEJ zone fracture toughness were made and compared.

Results: The DSP mouse had significantly harder enamel than the wild type. The DPP mouse had significantly softer, weaker, and poorly organized enamel. The Col1α2oim mouse had a significantly less tough DEJ zone than the wild type mouse; and more often underwent interfacial delamination.

Discussion: Over-expression of DSP throughout bulk enamel formation significantly increased hardness, consistent with its proposed role in creating harder damage diffusing specialized first formed enamel adjacent to the DEJ. Collagen has a structural role in toughening the DEJ and in preventing interfacial failure. We present a first "blueprint" of an engineered DEJ.

INTRODUCTION

Teeth are hierarchically organized structures consisting of two composite bioceramics, ectodermally derived enamel, and ectomesenchymally derived dentin. Stiff, hard, brittle enamel provides a wear- and acid-resistant masticatory surface. Enamel is formed by ameloblasts upon the mesenchymally derived dentin matrix which is formed by odontoblasts (Figure 1). This creates an embryologically, skeletally and biomechanically unique interface, the dentinoenamel junction or DEJ. In contrast to enamel, dentin is soft, flexible and tough, able to absorb energy and resist fracture. Thus the success of teeth is dependent upon the durable union of these dissimilar bioceramic tissues by the DEJ. However, interfaces between such dissimilar materials usually concentrate stresses and often delaminate. Surprisingly, the DEJ successfully transfers applied loads from enamel to dentin for a human lifetime of masticatory and parafunctional activities and occasional impact loads in a hostile warm wet and sometimes acidic environment. Enamel rarely separates from dentin. Furthermore, enamel cracks usually do not continue to propagate past the DEJ into dentin and through the entire tooth.[1,2]

The success of the DEJ has important clinical implications, as demonstrated by patients whose sound enamel delaminates in conditions such as dentinogenesis imperfecta. Also, the life-long success of the DEJ stands in stark contrast to the interfaces between artificial dental restorations and dentin. Artificial restorations such as fillings, crowns, and bridges typically fail either by bacterial invasion of this interface resulting in caries or pulpal death, or by gross structural failure resulting in dislodgement of the artificial restoration.[3-6] Hence, knowledge of structure-function relationships governing the DEJ would be of enormous therapeutic impact and allow better biomimetic interfaces between artificial restorations and dentin to be designed and eventually lead to a tissue-engineered DEJ.

Toughening mechanisms used to reduce crack propagation in engineered materials include: crack deflection; localized deformation; crack bridging by fibers, grains or other discontinuities; localized changes in elastic modulus or in microstructure, and crack blunting.[7] Such strategies may result in limited localized damage rather than in immediate total catastrophic failure of the entire structure. Some of these mechanisms may be related to the behavior of enamel and dentin in the DEJ zone.[8-35]

The outstanding mechanical success of the DEJ has been related to its anatomical structure including its high surface area (scalloping and microscalloping),[7] its collagen fiber content,[7,8] the mechanically irregular interface, intimate contact between enamel and dentin crystals,[8-10] and to a gradation of mechanical properties rather than an abrupt transition.[12] Although recent nano-indentation profiles and transmission electron microscopy demonstrate an abrupt transition between dentin and enamel at the DEJ interface,[11,13-15] wider micro-indentation profiling, micro-radiographic profiling, and Moire fringe analysis are suggestive of a broader functional zone. We propose that this functional zone contains mechanically specialized enamel and dentin adjacent to the interface.[16-18] Specialized highly mineralized first-formed enamel and specialized tubule-rich mantle dentin have long been known by microscopists to be present close to the DEJ.[19-21]

We have shown that the dominant failure mechanism of the human DEJ involves damage distribution within the specialized first formed enamel close to the DEJ interface.[35] Enamel cracks generally approach the DEJ obliquely within the specialized zone of harder first-formed enamel. Most enamel cracks are arrested at the DEJ interface, rarely crossing in to dentin. Many enamel cracks appeared to follow rod/interrod organization. Crack branching was also noted in enamel. Widespread enamel damage usually occurred in conjunction with small localized areas of interfacial delamination. Rare dentin cracks appeared to have a similar preference to run within the specialized zone of mantle dentin, rather than along the DEJ interface itself. Instead of continuous crack propagation, an interrupted crack occasionally occurred with multiple small discrete diagonal tears running from the DEJ interface into dentin, which is suggestive of a crack bridging mechanism in dentin. Thirdly, the DEJ interface often underwent plastic deformation without de-lamination. Overall, damage was spread over a broad zone, mostly first-formed enamel, rather than being localized at the DEJ interface. The specialized first-formed enamel may act as a shield to diffuse and spread damage so as to avoid catastrophic interfacial failure and spreading of superficial enamel cracks into bulk dentin.

Localized changes in tissue structure, e.g. in enamel, can only be reflections of alterations in the proteinaceous matrices that guide mineralization, which in turn are defined by gene expression (Figure 1). From a gene-expression perspective, many gene products of the dentin, and enamel, are "relatively" unique to those tooth matrices. For example, amelogenin, ameloblastin and enamelin are generally considered enamel-specific, while dentin-matrix protein-1 (DMP-1) and DSPP have generally been considered dentin-specific. As molecular probing techniques have become more sensitive, these gene expression pattern profiles have been challenged. In mice, the mRNAs of amelogenin, ameloblastin and enamelin, in addition to being a product of ameloblast cells, are also apparent in pre-secretory and secretory odontoblasts.[36-40] The mRNA and protein expression of dentin-matrix protein 1 (DMP-1) and DSPP, in addition to odontoblasts, are also apparent in pre-secretory ameloblasts.[41-45] This transient duplication of tooth-specific genes products during the early stages of dentin and enamel formation is suggestive of a molecular mechanism responsible for the creation of a transition zone at the DEJ.

Recent data suggest that the enamel proteins amelogenin, ameloblastin and enamelin have the biochemical ability to interact with a number of the collagens, including the type II (Col2a1) and the type V collagens (Col5a1 and Col5a3).[46] To date no evidence exists that would suggest that the ameloblasts transcribe any of the collagens. Collagen is thus considered a secreted product of cell types of mesenchymal origins only. These potential enamel protein-collagen interactions were identified using the yeast two-hybrid assay. If any of these enamel protein-collagen relationships prove to be physiologically valid, this would help to explain the DEJ as a transition zone, rather than an interface between the two dissimilar biological materials. It appears that no comprehensive cataloguing of the collagens in dentin has been done recently, however many collagens are expressed by odontoblasts including types I, III, IV, V[47] and VII (http://bite-it.helsinki.fi/).

Therefore, to gain insight into gene to mechanical function relationships, a series of transgenic mouse models, with known genetic changes in the genes thought to be involved in DEJ formation have been compared to wild-type mice using imaging, histological, and micromechanical testing modalities.

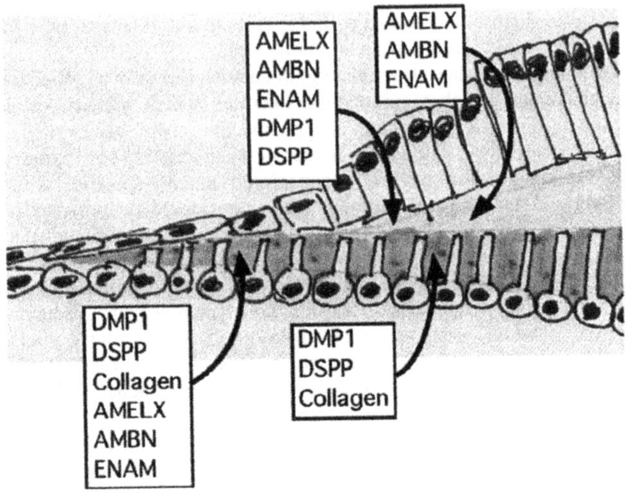

FIGURE 1. Early Events of Amelogenesis and Odontogenesis.
Dentin producing odontoblast cells transition from non-polar-non-matrix producing cells (left) to polar, matrix producing cells with processes extending into the newly formed dentin (shown as the orange material). Enamel producing ameloblast cells (top) transition from non-polar, non-matrix producing cells to polar, matrix producing cells. Enamel is shown as yellow. With the exception of the collagens, which are not a product of ameloblasts, X-chromosome derived amelogenin (AMELX), ameloblastin (AMBN), enamelin (ENAM), dentin matrix protein 1 (DMP1) and dentin sialophosphoprotein (DSPP) are all produced, and secreted from ameloblasts and odontoblasts as the dentin enamel junction is being established. After the establishment of the dentin enamel junction, the production of DMP1 and DSPP ceases in ameloblast cells, while the production of AMELX, AMBN and ENAM cease from odontoblasts.

MATERIAL AND METHODS

Transgenic Mice. It is important to note that we use transgenic mice with subtle alteration in function, so that the contribution of the gene and its protein can be identified and measured in a quasi-functional system. This is in contrast to knockout models, where a protein is completely removed and the resulting system is non-functional or absent. In our experiments, the transgenic mice were compared to age-matched wild type mice of similar genotype.

DSP Mice. It is now known that there is a transient expression of the DSPP gene in early ameloblasts adjacent to the DEJ, as well as in odontoblasts at all stages of development. Thus, the DSPP gene is expressed in the location of the DEJ adjacent first formed enamel, but not in bulk enamel. The DSPP gene encodes for multiple proteins, one of which is dentin sialoprotein or (DSP). The DSP mouse has a promoter that over-expresses the DSP protein throughout enamel formation.[48] We propose that this pattern of protein expression contributes to the formation and the remarkable mechanical properties of the DEJ.

DPP Mice. The DSPP gene also encodes for the dentin phosphoprotein protein, DPP. The DPP mouse has a promoter that over-expresses the DPP protein throughout enamel formation.[48]

OIM / Col1α2oim Mice. The Col1α2oim homotrimer mouse has collagen of the form $[\alpha 1(I)]_3]$, instead of the usual Type I collagen, $[\alpha 1(I)]_2[\alpha 2(I)]$, the dominant dentin protein, and the dominant structural collagen.[49] Should collagen contribute to the structural integrity of the DEJ, replacement of the primary structural collagen with a rare non-structural collagen would be expected to produce measurable mechanical changes.

Hardness and Fracture Toughness Indentations. Ten freshly extracted intact old lower left mouse incisors were extracted from each type of humanely sacrificed mice, and kept moist at all times and mounted in epoxy resin, were sequentially ground to a 0.1 μm alumina finish using a semiautomatic polisher (Buehler, Lake Bluff, IL). Vickers micro-indentations were made at 10 sites on bulk enamel, bulk dentin and at the DEJ for each specimen. Means and standard deviations were calculated for bulk enamel hardness (GPa), bulk enamel fracture toughness [K(IC), MPa.m(1/2)], bulk dentin hardness (GPa), and DEJ zone fracture toughness [K(IC), MPa.m(1/2)], were made using methods previously described,[18,35,50-53] The indentations were examined by optical microscopy, using polarization, interference, light/dark field, and transillumination techniques to identify cracks.

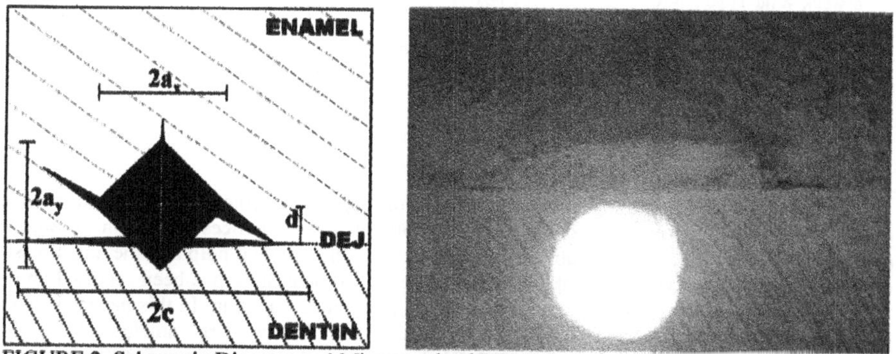

FIGURE 2. Schematic Diagram and Micrograph of DEJ Indentation.
Schematic diagram shows indentation ($2a_x$, $2a_y$) centered close to DEJ interface at distance d, and typical deflection of radial cracks by enamel rod microstructure as well as interfacial damage (2c). Light micrograph of similar indentation shows shadowing due to deflected radial cracks and interfacial delamination.

RESULTS

DSP Mice. No differences were recorded in dentin or at the DEJ zone where DSP is normally expressed. The microstructure if the DSP mouse enamel appeared to be the same as that of wild-type mice. However, over-expression of DSP throughout bulk enamel formation significantly increased hardness, compared to normal wild-type bulk enamel, consistent with its proposed role in creating harder damage diffusing first-formed enamel adjacent to the DEJ. Therefore, DSP has a structural role in creating specialized harder enamel.

DPP Mice. The quality, quantity, and microstructure of the enamel in these mice were profoundly inferior to those of wild-type mice. This suggests that DPP does not substantially contribute to the specialized hard first formed enamel. Possibly, DPP contributes to the specialized layer of last-formed outer enamel, which is believed to be softer and less organized than bulk enamel.

OIM / Col1α2oim Mice: These mice displayed a slight reduction in DEJ fracture toughness compared to wild-type mice. However, their DEJ interfaces underwent delamination at a frequency almost an order of magnitude more commonly than wild-type mice, 30% versus 3%. This suggests that collagen has an important structural role at the DEJ interface.

TABLE 1. Mechanical Properties of Transgenic and Age-Matched Wild-Type Mouse Teeth.

	Enamel Hardness	Enamel K(IC)	Dentin Hardness	DEJ zone K(IC)
4-week DSP	2.3 [0.3]	1.2 [0.2]	0.42 [0.04]	0.95 [0.1]
4-week DPP	1.7 [0.3]	0.9 [0.2]	0.42 [0.07]	0.53 [0.3]
4-week Wild Type	2.0 [0.2]	1.3 [0.1]	0.41 [0.04]	0.92 [0.1]
6-week Wild Type	NA	NA	0.52 [0.04]	1.11 [0.2]
6-week OIM	NA	NA	0.51 [0.07[	0.86 [0.2]

Means and standard deviations bulk enamel hardness (GPa), bulk enamel fracture toughness [K(IC), MPa.m(1/2)], bulk dentin hardness (GPa), and DEJ zone fracture toughness [K(IC), MPa.m(1/2)].

DISCUSSION

The DEJ is formed by a unique admix of expressed proteins contributed by the odontoblasts and ameloblasts lay down their early proteinaceous matrices and begin to migrate away from one another, the odontoblasts inwards towards the eventual pulp and the ameloblasts outwards towards the eventual tooth surface (Figure 1). To date, little has been reported about the contributions of any of the expressed proteins to mechanical function. In this paper we have demonstrated the importance of transient expression of DSP by early ameloblasts to the formation of specialized first-formed enamel. We have shown that expression of DPP is unlikely to contribute to the specialization of the first formed enamel. We have also demonstrated the importance of Type I collagen, $[\alpha 1(I)]_2[\alpha 2(I)]$, in resisting interfacial delamination of the DEJ. Hence, we present a first "blueprint" for the bioengineering of a DEJ.

Our long-term goal is to bioengineer DEJs. However, our short-term goal is to produce a superior artificial biomimetic junction between artificial restorations and dentin, a "BDRJ". This is important because despite use of highly evolved resinous adhesive dentin-bonding agents, artificial dental restorations typically fail at their junction with dentin. In the meantime, preservation of the natural DEJ during tooth preparation for artificial restorations is very strongly recommended, because in contrast to dentin, strong predictable and durable bonds can be achieved by the application of resinous adhesives to etched enamel.

We are unaware of other examples where a single gene encodes for proteins that appear to have directly opposing effects. The opposing effects of DSP and DPP suggest that post-transcriptional regulation may be important in producing specialized enamels.

Very recently, it has been shown that the porcine DSPP gene encodes for a third protein, dentin glycoprotein, DGP, or "the protein in the middle of the dentin sialophosphoprotein chimera.[54] Again, it is possible that this protein is also involved in producing specialized enamels.

Although, engineers often attempt to "make a better mousetrap", rarely is a better mouse engineered. We have engineered a mouse with harder enamel than the wild-type. Hardness is important because it is a surrogate for mineral density and is related to wear resistance and acid or caries resistance. Thus, the DSP mouse may have important therapeutic implications for new human therapies. However, there may be little evolutionary value in the wild type mouse having harder incisor teeth, because in comparison to humans, the mouse has a relatively short lifespan, continuously erupting and forming incisors, and lacks a cariogenic diet.

The OIM / Col1α2oim mouse is an established model for some forms of osteogenesis imperfecta (Brittle Bone Disease),[49,55] which is closely related to some forms of dentinogenesis imperfecta. In dentinogenesis imperfecta, healthy sound normal enamel delaminates from the supporting dentin. The exposed dentin is quickly worn away or damaged by caries. This has tremendous functional, self-image, treatment, and cost implications for affected people. In our study, the hardness of the affected dentin did not differ from the control wild-type teeth. Because hardness is known to be affected by elastic modulus, our mouse microindentation data suggests that the increased delamination in the OIM / Col1α2oim mouse teeth was caused by an interfacial weakness. However, an alternative or additional explanation for enamel loss in affected human teeth might be that a reduced dentin elastic modulus, or a lower proportional limit, might allow increased flexural stresses to be applied to brittle enamel, also contributing to its fracture, delamination, and loss. Direct measurement of dentin elastic modulus could clarify this issue.

Collagen is a likely candidate for crack bridging both at the DEJ interface and within bulk dentin. The results of this experiment affirm that collagen plays a structural role in preventing DEJ interfacial delamination. Collagen might not only resist crack initiation at the DEJ, but might also limit interfacial crack propagation. Collagen bridging may account for the interupted cracking mode that has been recently reported in dentin.[33,35]

This microindentation study applied high pressures, high stress intensities, single quasi-static loads, short cracks, and produced substantial amounts of damage within the DEJ zone. Thus our experimental method may not be a realistic model of physiologically occurring DEJ zonal damage, which includes millions of repetitive low stress cycles and occasional impact loads. Furthermore, the mouse is not a completely accurate model of human enamel microstructure and may not be a completely accurate model of human protein expression. However, due to our ability to make prospective changes, defined genetics, low variance, and quickly achieved maturity, the transgenic mouse remains an invaluable investigative method.

CONCLUSION

A novel first blueprint for the bioengineering of the dentino-enamel junction has been presented. Regulated gene expression of Type I Collagen and Dentin Sialoprotein contributes to the unique crack resistant properties of the dentino-enamel junction.

ACKNOWLEDGEMENT

The authors most gratefully acknowledge financial support form the National Institutes of Health / National Institution for Dental and Craniofacial Research under the following grants for this work: NIH/NIDCR DE14189, DE13045, DE13404. We are also most grateful for the gracious and collegial gift of knowledge, specifically the cDNA mouse DSPP, from Mary MacDougall of the University of Alabama at Birmingham.

REFERENCES

[1]Zachrison BU, Skogan O, Hoymyhr S. Enamel cracks in debonded, debanded, and orthodontically untreated teeth. *Am J Orth* 1980;77:307-19.

[2]Walker BN, Makinson OF, Peters MC. The role of enamel lamellae in caries initiation. *Aust Dent J* 1998;43:110-6.

[3]Brannstrom M, Nyborg H. Bacterial growth and pulpal changes under inlays cemented with zinc phosphate cement and epoxylite CBA 9080. J Prosthet Dent 1972;31:556-65.

[4]Coornaert J, Adriaens O, De Boever J. Long term study of porcelain-fused-to-gold restorations. *J Prosthet Dent* 1984;51:338-42.

[5]Glantz PO, Ryge G, Jendressen MD, Nilner K. Quality of extensive fixed prosthodontics after five years. *J Prosthet Dent* 1984;52:474-9.

[6]Leempoel PJ, Eschen S, De Haan AF, Van't Hof M.A. An evaluation of crowns and bridges in a general dental practice. *J Oral Rehabil* 1985;12:515-28.

[7]Lawn B. Fracture of brittle solids. Second Edition. Chapter 7, Microstructure and toughness. Cambridge: Cambridge University Press; 1993. p194-248.

[8]Lin CP, Douglas WH, Erlandsen S.L. Scanning electron microscopy of Type I collagen at the dentin enamel junction of human teeth. *J Histochem Cytochem* 1993;41:381-8.

[9]Arsenault AL, Robinson BW. The dentino-enamel junction: A structural and microanalytical study of early mineralization. *Calcif Tissue Int* 45;1989:111-21.

[10]Hayashi Y. High resolution electron microscopy in the dentino-enamel junction. *J Electron Microsc* 1992;41:387-91.

[11]Fong H, Sarikaya M, White SN, Snead ML. Nano-mechanical properties profiles across dentin enamel junction of human incisor teeth. *Mat Sci Eng C* 2000;7:119-28.

[12]Lin CP, Douglas WH. Structure-property relations and crack resistance at the bovine dentin-enamel junction. *J Dent Res*, 1994;73:1072-1078.

[13]Marshall GW Jr, Balooch M, Gallagher RR, Gansky SA, Marshall SJ. Mechanical properties of the dentinoenamel junction: AFM studies of nanohardness, elastic modulus and fracture. *J Biomed Mater Res* 2001;54:87-95.

[14]Habelitz S, Marshall SJ, Marshall GW, Balooch M. The functional width of the dentino-enamel junction determined by AFM-based nanoscratching. *J Struc Biol* 2001; 135: 294-301.

[15]Balooch G, Marshall GW, Marshall SJ, Warren OL, Asif SAS, Balooch M. Evaluation of a new modulus mapping technique to investigate microstructural features of human teeth. *J Biomech* 2004;37:1223-32.

[16]Wang RZ, Weiner S. Strain-structure relations in human teeth using Moire fringes. *J Biomech* 1988;31:135-41.

[17]Elliott JC, Anderson P, Gao XJ, Wong FSL, Davis GR, Dowker SEP. Application of scanning microradiography and X-ray microtomography to studies of bones and teeth. *J X-Ray Sci. Technol* 1994;4:102-17.

[18]White SN, Paine ML, Luo W, Sarikaya M, Fong H, Yu Z, Li ZC, Snead ML. The dentino-enamel junction is a broad transitional zone uniting dissimilar bioceramic composites. *J Am Ceram Soc* 2000;83:238-40.

[19]von Ebner V. Uber die histologischen veranderungen des zahnschmelzes wahrend der erhartung, inbesondere beim menschen. *Arch Mikrosk Anat* 1906;67:18-81.

[20]von Korff K. Die entwicklung der zahnbeingrundsubstanz der saugetiere. *Arch Mikrosk Anal* 1906;67:1-17.

[21]Weidenreich F. Uber den bau und die entwicklung des zahnbeins in der reihe der wirbeltiere. *Anat Entw Gersch* 1925;76:218-60.

[22]Rasmussen ST, Patchin RE, Scott DB, Heuer AH. Fracture properties of human enamel and dentin. *J Dent Res* 1976;55:154-64.

[23]Hassan R, Caputo AA, Bunshaw RF. Fracture toughness of Human Enamel. *J Dent Res* 1981;60:820-827.

[24]Rasmussen ST. Fracture properties of human teeth in proximity to the dentinoenamel junction. *J Dent Res* 1984;63:1279-83.

[25]El Mowfay OM, Watts DC. Fracture toughness of human dentin. J Dent Res 1986;65:677-81.

[26]Pioch T, Golfels D, Staehle HJ. An experimental study of the stability of irradiated teeth in the region of the dentinoenamel junction. *Endod Dent Traumatol* 1992;8:241-4.

[27]Pioch T, Staehle HJ. Experimental investigation of the shear strengths of teeth in the region of the dentinoenamel junction. *Quintessence Int* 1996;27:711-714.

[28]Mjor I A, Nordahl I. The density and branching of dentinal tubules in human teeth. *Archs Oral Biol* 1996;41:401-12.

[29]Marshall GW, Marshall SJ, Kinney JH, Balooch M. The dentin substrate: Structure and properties related to bonding. *J Dent* 1997;25:441-58.

[30]Xu HH, Smith DT, Jahanmir S, Romberg E, Kelly JR, Thompson VP, Rekow ED. Indentation damage and mechanical properties of human enamel and dentin. *J Dent Res* 1998;77:472-80.

[31]White SN, Luo W, Paine ML, Fong H, Sarikaya M, Snead ML. Biological organization of hydroxyapatite crystallites into a fibrous continuum toughens and controls anisotropy in human enamel. *J Dent Res* 2001;80:321-6.

[32]Cuy JL, Mann AB, Livi KJ, Teaford MF, Weihs TP. Nanoindentation mapping of the mechanical properties of human molar tooth enamel. Arch Oral Biol 2002; 47: 281-291.

[33]Kahler B, Swain MV, Moule A. Fracture-toughening mechanisms responsible for differences in work to fracture of hydrated and dehydrated dentine. *J Biomech* 2003;36:229-37.

[34]Dong XD, Ruse ND. Fatigue crack propagation path across the dentinoenamel junction complex in human teeth. *J Biomed Mater Res* 2003;66A:103-9.

[35]White SN, Miklus VG, Chang PP, Caputo AA, Fong H, Sarikaya M, Luo W, Paine ML, Snead ML. Controlled failure mechanisms toughen the dentino-enamel junctional zone.*J Prosthet Dent* 2005 In Press.

[36]Begue-Kirn C, Krebsbach PH, Bartlett JD, Butler WT. Dentin sialoprotein, dentin phosphoprotein, enamelysin and ameloblastin: tooth specific molecules that are distinctively expressed during murine dental differentiation. Eur J Oral Sci 1998;106:963-70.

[37]Oida S, Nagano T, Yamakosh Y, Ando H, Yamada M, Fukae M. Amelogenin gene expression in porcine odontoblasts. *J Dent Res* 2002;81:103-8.

[38]Papagerakis P, MacDougall M, Berdal A. Differential epithelial and mesenchymal regulation of tooth-specific matrix proteins expression by 1,25-dihydroxyvitamin D3 in vivo. *Connect. Tissue Res* 2002;43:372-5.

[39]Nagano T, Oida S, Ando H, Gomi K, Arai T, Fukae M. Relative levels of mRNA encoding enamel proteins in enamel organ epithelia and odontoblasts. *J. Dent. Res*. 2003;82:982-6.

[40]Lu P, Gao XJ, Jia HT, Wang JD, Gao Y, Wei MJ. Enamelin transcriptional expression in developing postnatal rat tooth germ. *Zhonghua Kou Qiang Yi Xue Za Zhi* 2004;39:414-7.

[41]MacDougall M, Nydegger J, Gu TT, Simmons D, Luan X, Cavender A, D'Souza RN. Developmental regulation of dentin sialophosphoprotein during ameloblast differentiation: a potential enamel matrix nucleator. *Connect Tissue Res* 1998;39:25-37.

[42]Ritchie HH, Berry JE, Somerman MJ, Hanks CT, Bronckers AL, Hotton D, Papagerakis P, Berdal A, Butler WT. Dentin sialoprotein (DSP) transcripts: developmentally-sustained expression in odontoblasts and transient expression in pre-ameloblasts. *Eur J Oral Sci* 1997;105(5 Pt 1):405-13.

[43]D'Souza RN, Cavender A, Sunavala G, Alvarez J, Ohshima T, Kulkarni AB, MacDougall M. Gene expression patterns of murine dentin matrix protein 1 (Dmp1) and dentin sialophosphoprotein (DSPP) suggest distinct developmental functions in vivo. *J Bone Miner Res* 1997;12:2040-9.

[44]Begue-Kirn C, Ruch JV, Ridall AL, Butler WT. Comparative analysis of mouse DSP and DPP expression in odontoblasts, preameloblasts, and experimentally induced odontoblast-like cells. *Eur J Oral Sci* 1998;106 (Suppl. 1): 254-9.

[45]Baba O, Qin C, Brunn JC, Jones JE, Wygant JN, McIntyre BW, Butler WT. Detection of dentin sialoprotein in rat periodontium. *Eur J Oral Sci* 2004;112:163-70.

[46]Wang HJ, Tannukit S, Zhu DH, Snead ML, Paine ML. Enamel matrix protein interactions. J Bone Miner Res. 2005;20:1032-40.

[47]Bronckers AL, Gay S, Lyaruu DM, Gay RE, Miller EJ. Localization of type V collagen with monoclonal antibodies in developing dental and peridental tissues of the rat and hamster. *Coll Rela. Res* 1986;6:1-13.

[48]Paine ML Paine ML, Luo W, Wanh H-J, Ngan AYW, Miklus VG, Zhu D-H, MacDougall M, White SN, Snead ML. Dentin sialoprotein and dentin phosphoprotein over-expression during amelogenesis. *J Biol Chem* 2005; In Press.

[49]Miles CA, Sims TJ, Camacho NP, Bailey AJ. The role of the α2 chain in the stabilization of the collagen type I homotrimer in *oim* mouse tissues. *J Mol Biol* 2002;321:797-805.

[50]Zhang J, Lewandowski JJ. Interfacial fracture toughness measurement using indentation. *J Mat Sci* 1994;29:4022-6.

[51]Evans AG, Charles EA. Fracture toughness determinations by indentation. *J Am Ceram Soc* 1976;59:371-2.

[52]Anstis GR, Chantikul P, Lawn BR, Marshall DB. A critical evaluation of indentation techniques for measuring fracture toughness. *J Am Ceram Soc* 1981;64;533-8.

[53]Wuttiphan S, Lawn BR. Crack supression in strongly bonded homogenous/heterogenous laminates: A study on glass/glass-ceramic bilayers. *J Am Ceram Soc* 1996;79:634-40.

[54]Yamakoshi Y, Hu J C-C, Fukae M, Zhang H, Simmer JP. Dentin glycoprotein, the protein in the middle of the dentin sialophosphoprotein chimera. *J Biol Chem* 2005;280:17472-9.

[55]Phillips CL, Bradley DA, Schlotzhauer CL, Bergfeld M, Liberos-Minotta C, Gawenis LR, Morris JS, Clarke LL, Hillman LS. Oim mice exgibit altered femur and incisor mineral composition and decreased bone mineral density. *Bone* 2000;27:219-26.

Characterization of Interfaces

ABLATION IN CARBON/CARBON COMPOSITES: MICROSCOPIC OBSERVATIONS AND 3D NUMERICAL SIMULATION OF SURFACE ROUGHNESS EVOLUTION

J. Lachaud[1], G. L. Vignoles[1], J.-M. Goyhénèche[1], J.-F. Epherre[1,2]

[1]Laboratoire des Composites ThermoStructuraux (LCTS),
UMR 5801: CNRS-SAFRAN-CEA-UB1,
Domaine Universitaire de Bordeaux – 3, Allée de La Boétie, 33600 Pessac, France

[2]Commissariat à l'Energie Atomique (CEA), CESTA, BP2, 33114 Le Barp, France

ABSTRACT

The design of efficient thermal protection systems relies on the use of *ablative* thermostructural materials with controlled surface recession, such as Carbon/Carbon (C/C) composites. Hence the coupled physical phenomena leading to carbon loss and resulting in a global surface recession have to be known. Here, the focus is set on a multi-scale three-dimensional study of appearing composite surface roughness which, directly or not, induces a strong enhancement of heat and mass transfer and consequently increases ablation velocity.

First, surface roughness of C/C samples submitted to arc-jet ground test (quite representative of real conditions) is mainly investigated by scanning electron microscopy (SEM). A description, a classification and an analysis of multi-scale surface roughness features are proposed.

Then a three-dimensional reaction-diffusion local model is set up to explain the formation of the typical needle shape of the carbon fibers in emerging yarns. A 3D numerical code using Brownian motion simulation technique (gas diffusion) and random sticking events (heterogeneous reaction) has been implemented to solve a diffusion-reaction model. The surface is described with a simplified Marching Cubes approach, which allows to handle efficiently its alteration by ablation. The simulation method is shown to give, in a reasonable computing time, results in correct agreement with SEM observations.

Moreover, such an approach could contribute to the identification of material or flow intrinsic properties through morphological analysis of the roughness features.

INTRODUCTION

Spatial vehicles often encounter severe environments during atmospheric crossing. Since the speed can reach many km/s in dense atmosphere, the flow close to the nosetip can be hypersonic. In this case a thin bow shock surrounds an inviscid flow (eulerian flow), while a dynamic boundary layer develops in the vicinity of the wall[1] (see left part of Fig. 1). Typically, temperature of the flow may reach $7000\,K$ and it may lead to a maximal wall temperature of $4500\,K$ for pressures higher than $100\,bar$.[2] For such an environment, the design of thermal protection systems relies on Carbon/Carbon (C/C) composites, which possess the best compromise between thermal, thermo-chemical and mechanical properties.[3] However, a high interfacial mass transfer, strongly coupled with boundary layer transfer phenomena, leads to surface recession. Indeed, C/C composites are progressively destroyed by oxidation, sublimation and, up to a certain extent, mechanical erosion. These physico-chemical phenomena, denoted by the term *ablation*, are globally endothermic. Hence they reduce the wall

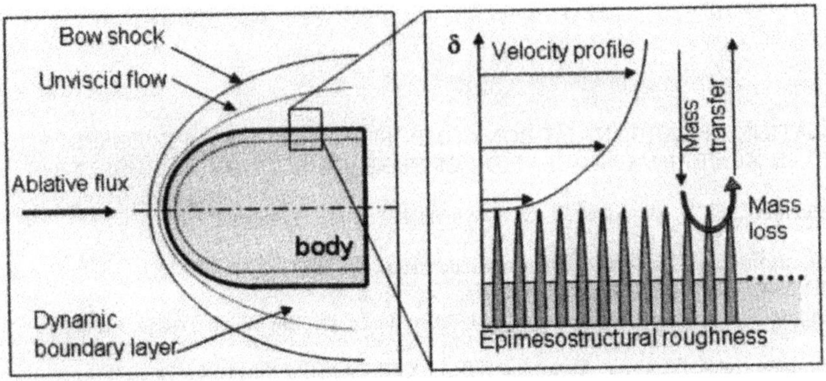

Figure 1: Hypersonic flow over a blunt body at macroscopic (left) and yarn (right) scales.

temperature and the heat flux that penetrates the internal structure.

Ablation of C/C composites quickly leads to surface roughness. Roughness induces an enhancement of heat and mass transfer which puts up ablation velocity via two major phenomena: (i) it increases the chemically active surface of the wall; (ii) it contributes to the laminar-to-turbulent transition in the dynamic boundary layer. If general phenomenological tendencies are predictable, the understanding of the interaction between the flow and the material has to be improved. In this work, an effort is done to improve this understanding in laminar flow conditions through the observation, the study and the modelling of roughness evolution. This document includes four parts:

1. First, a description and a classification of multi-scale surface roughness features appearing on C/C composites are proposed;
2. Then, a three-dimensional reaction-diffusion local model is set up to explain the formation of the typical needle shape of emerging carbon fibers;
3. A homemade 3D numerical simulation code, called AMA, developed to solve moving interface problem coupled to "heterogeneous reaction/ mass transfer" processes is succinctly presented;
4. Last, the model established in part 2 is solved using AMA. Results, which are in correct agreement with observations, enable optimistic outlooks for the improvement of ablation understanding, and the determination of material intrinsic properties.

ROUGHNESS DESCRIPTION AND CLASSIFICATION

Test and observation conditions

Unfortunately, it is quite difficult to recover samples from real flight experiments. As far as roughness is concerned, arc-jet tests are supposed to be representative of real flight

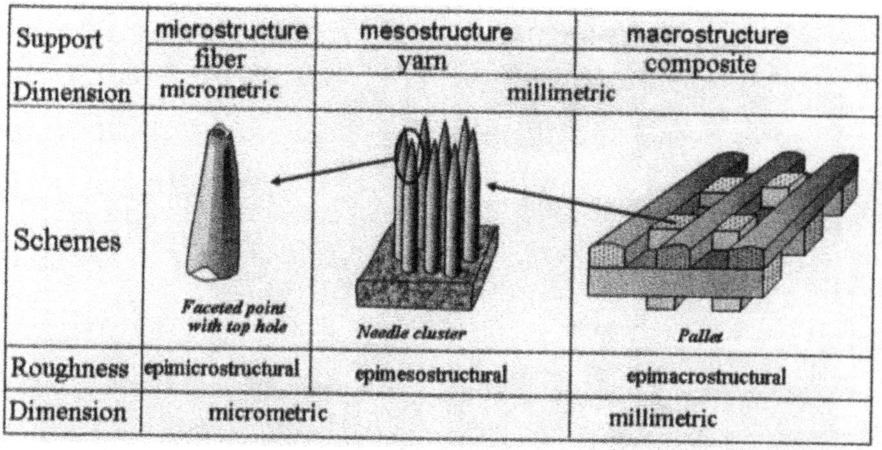

Figure 2: Multi-scale roughness on a 3D C/C ablated material : synthesis tabular.

conditions. This study is restricted to arc-jet tests; however its results may have some interest for real flight ablation simulation.

C/C composite samples, made from a 3D ex-PAN carbon fiber woven preform and a pitch-based carbon matrix, has been submitted to arc-jet ground tests in stagnation point configuration. The material temperature was high enough (3000 K) to enable both oxidation and sublimation. However, the efficient 3D C/C structure prevents the sample from being notably eroded during ablation. Surface roughness has been observed by binocular magnifier (BM), optical microscope (OM) and scanning electron microscopy (SEM).

Description and classification

As shown on the left sketch of the Fig. 2, 3D C/C is an heterogeneous multi-scale material, on which roughness features can appear during ablation. The small points represent the fiber sections (microstructure). Several thousands of fibers are linked together into yarn with a pitch-based matrix (mesostructure). Then, yarns are orthogonally fit together into a pattern repeated by translation on a cubic lattice. This macro-structure leads to a network of parallelepipedic macro-pores (located near each node of the lattice), which are partially filled with pitch matrix.

A classification of the observed roughness features according to their length scales is presented on Fig. 2. For each scale, a sketch and a name are given:

1. Epimacrostructural roughness takes place on the lattice. It seems to result from the difference of reactivity between yarns and extra-yarn pitch-based matrix. Mechanical erosion sporadically occurs through the detachment of an extra-yarn matrix octet. The section of emerging yarns (tangent or perpendicular to surface) is slightly undulating. Indeed, if edges of initially square section of yarns are emerging, creating notches, they

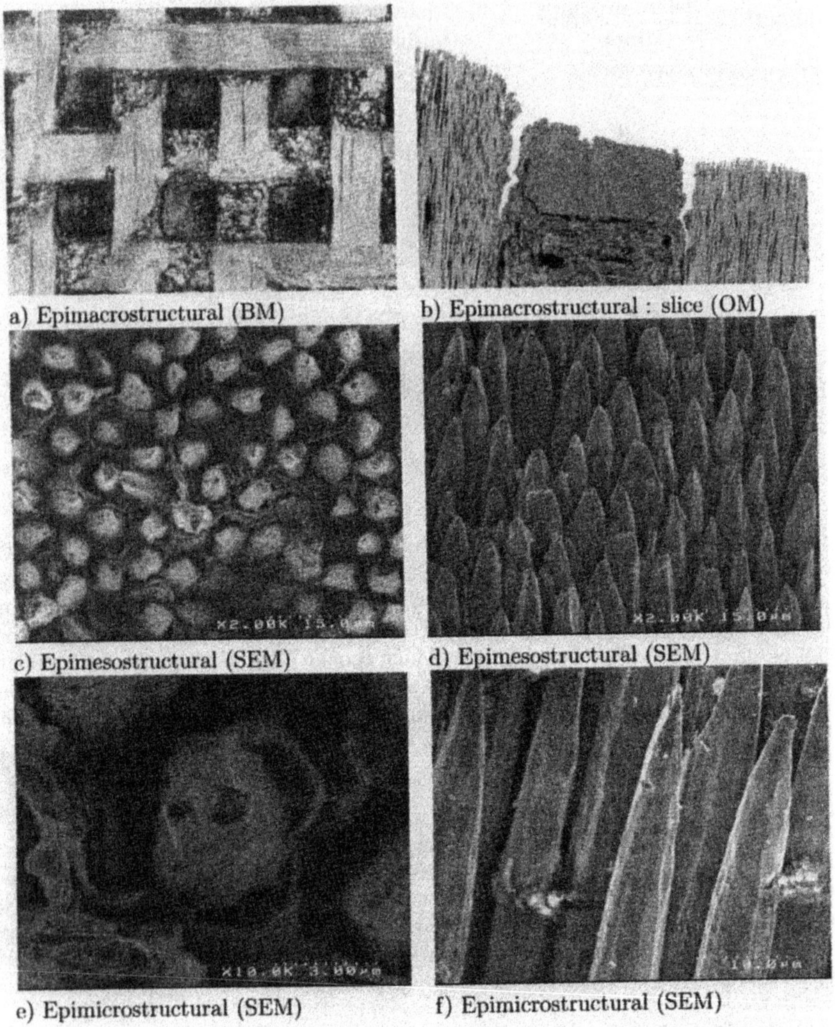

a) Epimacrostructural (BM) b) Epimacrostructural : slice (OM)

c) Epimesostructural (SEM) d) Epimesostructural (SEM)

e) Epimicrostructural (SEM) f) Epimicrostructural (SEM)

Figure 3: Micrographies of 3D C/C surface roughness after ablation.

are smoothed out to a wavy form by ablation. A BM photograph of the surface and a OM micrograph of a polished slice, presented on Fig. 3-a and b, associated to the knowledge of the material structure, enable us to sketch the "pallet" scheme.

2. Epimesostructural roughness develops at the end of emerging yarns, and looks like "needle clusters" (resp. "needle layers") for yarns perpendicular (resp. parallel) to material surface. In the literature, many micrographs show such roughness features on carbon-based composites during ablation by oxidation[4,5] or both oxidation and sublimation.[6–8] As shown on photographs 3-c and d, due to an important recession of the intra-yarn matrix, fibers, which are less reactive, are partially stripped, become thinner, and acquire a needle shape.

3. Epimicrostructural roughness appears on the microstructure. Fiber tips are faceted (Fig. 3-f). Moreover, Fig. 3-c and e also shows holes on the top of the fibers.

According to the classification presented above, ablation phenomena exhibits the material structure through roughness set-up. Geometrical features mainly follow the material structure. Then, let us call it structural roughness to make a difference with a purely physical roughness which has already been observed on homogeneous material and modelled.[2] This physical roughness consists in scalloped morphologies and is not correlated to material structure. The cause seems to be a dynamical effect based on the concurrence between mass transfer and chemical reactions.

Regarding structural roughness, physical phenomena are also underlying (reactivity differences and possibly perturbations of local mass transfer). As a result, models including structure and physics has to be taken in consideration.

MODEL SET-UP

Position of the study

In the literature, the global phenomenology of ablation at macroscopic scale is widely described from the fluid point of view, i.e. the material is considered as a reactive and thermally conductive homogeneous interface).[1,9–15] However, as the scope of this study is to understand structural roughness set-up, the problem is approached here from the composite material point of view. Moreover the choice is made to focus on the mesostructure scale ("needle clusters") and to study the stationary profiles.

Phenomenology

The global phenomenology has roughly been described in the introduction of this document. Let us now complete it with the yarn scale specific phenomenology. It is linked to macroscopic scale on Fig. 1. As far as reactivity is considered, the C/C composite yarns are heterogeneous. Indeed, for both oxidation (Eq. 1) and sublimation (Eq. 2),

$$C_{(s)} + O_2 \rightleftharpoons CO_2 \quad (1)$$

$$n\, C_{(s)} \rightleftharpoons C_n \quad (2)$$

the reaction rates of fibers (more organized turbostratic carbon[3,16]) are lower than the matrix ones. Mass loss is strongly coupled to mass transfer perpendicularly to the wall.[2] In

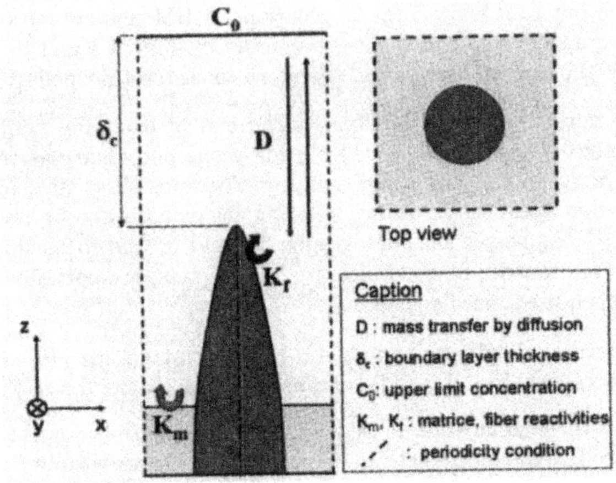

Figure 4: Scheme of the elementary pattern and of the proposed model

laminar flow conditions, mass transfer is mainly diffusive and can be described by a boundary layer formalism. Laminar boundary layer thickness (δ) has been estimated by Navier-Stokes equations resolution, its value is of the order of one millimeter. For each species i, it is reasonable to deduce mass concentration boundary layer thickness (δ_{c_i}) from δ by the following relation:[17]

$$\frac{\delta}{\delta_{c_i}} \approx Sc^{\frac{1}{3}} \tag{3}$$

where $Sc = \nu / D_{iM}$ is the Schmidt number (with ν and D_{iM} ($m^2 s^{-1}$) resp. kinematic viscosity and binary mass diffusion coefficient of species i in the gas mixture M), close to unity in gases.

The difference of reactivity between fibers and matrix leads to a local difference of ablation velocity which turns into surface roughness, unless gas diffusion phenomena brings a limit (see part 4).

Composite surface model and conditions

The proposed model is sketched on Fig. 4. On this scheme the stationary roughness is represented; however at initial time, the fluid/solid interface is flat. This profile has therefore to be obtained using a moving interface modelling.

The model includes the following hypotheses:

- Material:
  - The yarn is made of fibers and matrix;
  - Matrix and fibers are assumed homogeneous and isotropic;

- The composite is supposed to be only ablated by a first order oxidation reaction* (represented by Eq. 1); that is to say temperature is rather close to 3000K than higher;
- Yarn section is supposed infinite in transverse directions;

• Physical phenomena:

- Mass transfer is restricted to binary diffusion of the single reactant (O_2) in nitrogen (diffusion of the gaseous by products (CO_2) is neglected);
- Chemical reaction is restricted to the surface;
- Close to the wall, advection is negligible;
- Thermal gradients are neglected ;
- Stefan flux (due to mass transfer into the fluid) is neglected.

Let us write mass conservation of the reactant (of molar concentration C) in the fluid phase:

$$\frac{\partial C}{\partial t} + \nabla \cdot (-D\nabla C) = 0 \tag{4}$$

Relative boundary conditions are:

- On boundary layer top: $C = C_0$;
- At the fluid/solid interface : $(-D\nabla C) \cdot \mathbf{n} = -k_i C$ (where $\mathbf{n}$ is the normal to the surface, and k_i (m/s) the reaction kinetic constant of matrix ($i = m$) or fibers ($i = f$));
- Periodicity on the lateral boundaries.

The interface position ($z = h(x, y, t)$) is given by the coupled resolution of : $\frac{\partial h}{\partial t} = -v_s k_i C\mathbf{n}$ (where v_s is the solid molar volume). Moreover, δ_c is a constant, as the top of the boundary limit moves together with the wall.

DIRECT NUMERICAL SIMULATION IMPLEMENTATION AND VALIDATION

Numerical implementation

To solve this problem in 3-D and in non-stationary regime, a numerical simulation code, named AMA, has been developed on a Monte-Carlo random-walk principle. As sketched on Fig. 5, AMA, which is a C ANSI implementation, falls into five main parts:

- A 3-D image containing several phases (fluid/solids) is described by discrete cubic voxels method;
- The moving fluid/solid interface is determined by a simplified marching cube approach;[18]

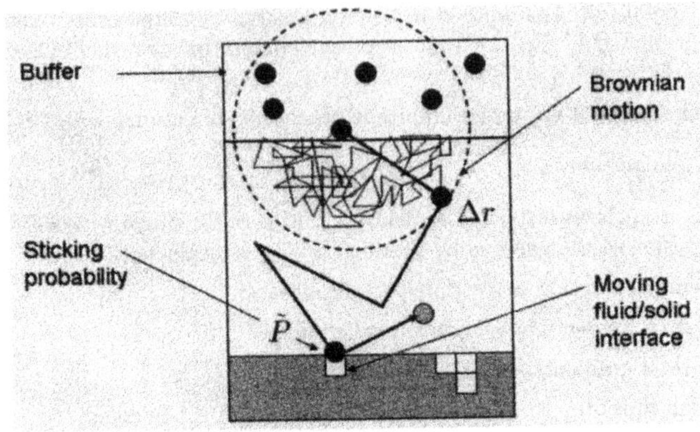

Figure 5: Scheme of the numerical simulation code AMA principle.

- Mass transfer by diffusion is simulated by a Brownian motion simulation technique,[19] which is a continuum (grid-free) and fast convergence method for diffusion in a continuous fluid. The time increment (Δt) is proportional to the square of space increment (Δr) and to the inverse of the binary diffusion coefficient (D).[20]
- Heterogeneous first order reaction on the wall is simulated by a sticking probability ($\tilde{P}$) adapted to the Brownian motion simulation technique.[21] The relation 5 has been deduced by a theoretical analysis. The value of α (which is close to 0.8) is obtained by a numerical study.

$$\tilde{P} = \frac{1}{1 + \frac{3/2\alpha D}{k\Delta r}} \tag{5}$$

- A Dirichlet upper boundary condition is simulated using a buffer.

AMA validation

AMA has been successfully compared to the 1-D (with i restricted to m) steady state diffusion/first-order reaction model reduced from the model proposed above. By solving equation 4, one obtains the analytical reactant flux at the interface (Φ):

$$\Phi = \frac{D}{\delta_c} C_0 \frac{1}{1 + \frac{1}{Da}} \tag{6}$$

where $Da = k_m \delta_c / D$ is a Damköhler number.

Fig. 6 illustrates the comparison between numerical simulation and analytical expression of Φ (cf. Eq. 6) when k_m varies. The maximal relative error is lower than 2.5%, whatever the values chosen for δ_c, C_0 and D.

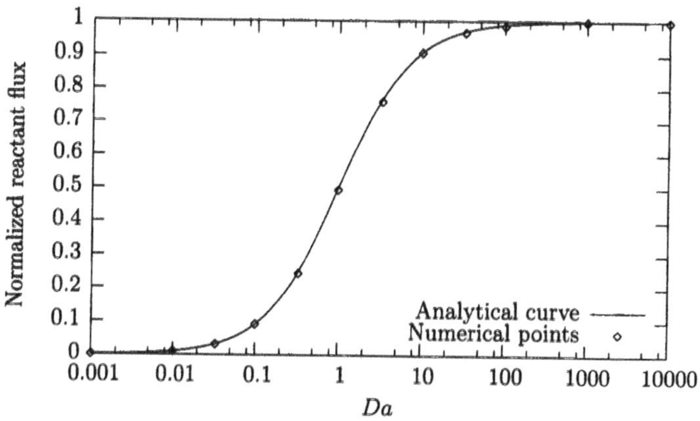

Figure 6: AMA validation by comparison to a 1-D model

NUMERICAL SIMULATION RESULTS

The lack of experimental data on turbostratic carbon reactivity to oxygen at high temperatures prevent us from doing a single and accurate simulation case. Moreover, the variety of turbostratic carbons is very large : reactivity may span as much as 3 orders of magnitude.[16] Consequently, it is more appropriate to perform a parametric study, using the model sketched on Fig. 4. The results are presented on Fig. 7. The influence of three dimensionless groups on the stationary roughness profiles of "needle clusters" is studied:

- Damköhler number for the fiber : $Da_f = k_f \delta_c / D$ (only k_f varies) (minimal value (min.): 0 , maximal value (max.): ∞);
- Reactivity ratio : $\tilde{k} = k_m / k_f$ (min.: 1 , max.: ∞);
- Fiber/cell ratio : $\tilde{r}_f = d_f / l$ (min.: 0 , max.: 1), with d_f: fiber diameter, and l: square base cell size.

The model sensitivity to each-one of these three dimensionless groups is tested from a central case ($Da_f = 0.1$, $\tilde{k} = 10$, $\tilde{r}_f = 0.6$).

For each simulation case, the steady state regime is reached after a computing time ranges from 1 to 24 hours with a Xenon CPU 3.2GHz processor.

Let us describe the morphological evolutions of roughness features as the dimensionless groups are varying. Peak-to-valley roughness increases together with:

- Fiber/cell ratio (r_f) in an almost homothetic way;
- The inverse of Damköhler number (Da_f) from a flat to an almost pyramidal geometry;
- Reactivity ratio ($\tilde{k}$) from a flat to a needle-like geometry.

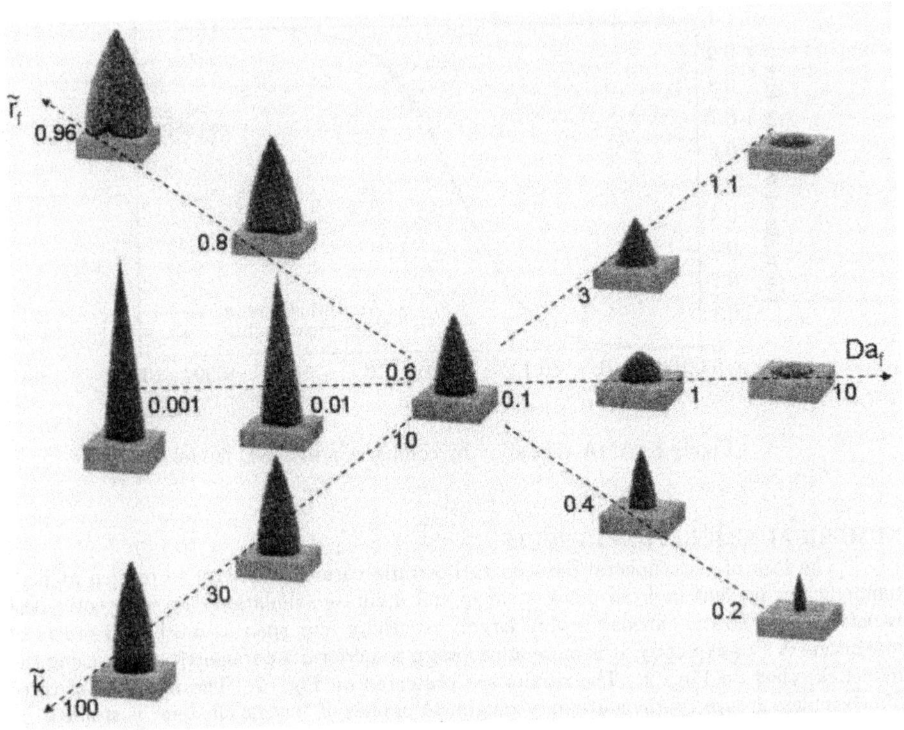

Figure 7: Parametrical study at mesoscopic scale

Limiting cases are observed. On the right side of the Fig. 7, roughness set-up becomes impossible, since the problem turns out to be 1-D for at least one of these reasons (from top to bottom): (i) homogeneous solid ($\tilde{k} \rightarrow 1$); (ii) diffusive limitation ($Da \rightarrow \infty$); (iii) homogeneous solid ($\tilde{r}_f \rightarrow 0$). On the left side, the peak-to-valley roughness is maximal since (from top to bottom): (i) fiber diameter is maximal ($\tilde{r}_f \rightarrow 1$), (ii) reactive limitation on fiber and matrix is reached ($Da_f \rightarrow 0$), (iii) diffusive limitation on matrix is reached ($\tilde{k} \rightarrow \infty$).

As a result, it is possible to analyze real morphologies through those first results. Of course, the simplicity of the model used has to be kept in mind. A comparison of Fig. 7 to micrographs of Fig. 3-d and f tends to show that the actual material reactivity ratio between fiber and matrix is high. Moreover, as fibers tips seem more rounded on micrograph 3-d, fibers are unlikely to be in reactive limitation. Nothing can be said on the matrix behavior.

It is reasonable to conclude that this simple results can qualitatively explain the development of "needle clusters" roughness features and contributes to understand ablation.

CONCLUSION AND OUTLOOK

In this work, the origin and development of roughness on a 3D-C/C ablative composite during atmospheric crossing are tackled. First, using as reference arc-jet test samples, multi-scale roughness is observed, classified and briefly analyzed as a whole. Then the focus is set on well-know "needle clusters" roughness features which develop on perpendicularly emerging yarns. A simple heterogeneous reaction/diffusion model is proposed. The use of an efficient homemade 3D numerical simulation code (validated by comparison to a 1-D analytical model) enabled to perform a parametrical study. Despite the lack of experimental data for 3D validation, results are promising, since the proposed model can qualitatively explain roughness development. Moreover, it is likely to give pertinent information on reactive regime conditions and helps us to improve the understanding of the role of coupled physicochemical phenomena in ablation.

Up to a certain extent, this study opens the possibility to determine, by inverse analysis, either mass transfer properties or material intrinsic reactivities.

FOOTNOTE

*Sublimation modelling is quite similar and gives the same stationary profiles since dimensionless groups values are equal.

ACKNOWLEDGEMENTS

The authors wish to thank CNRS and CEA for Ph. D. grant to J. Lachaud and Y. Aspa (LCTS) for fruitful discussions.

References

[1] J. Couzi, J. de Winne, and B. Leroy. Improvements in ablation predictions for reentry vehicle nosetip. In *Proceedings of the third European symposium on aerothermodynamics for space vehicles*, pages 493–499, Noordwijk, The Nederlands, 24-26 November 1998.

[2] G. Duffa, G. L. Vignoles, J.-M. Goyhénèche, and Y. Aspa. Ablation of C/C composites : investigation of roughness set-up from heterogeneous reactions. *International Journal of Heat and Mass Transfer*, 48(16):3387–3401, June 2005.

[3] L. M. Manocha and E. Fitzer. *Carbon reinforcement and C/C composites*. Springer, 1998.

[4] D. Cho, J. Y. Lee, and D. I. Yoon. Microscopic observations of the ablation behaviours of carbon fibre/phenolic composites. *Journal of materials science*, 12:1894–1896, 1993.

[5] E. Duvivier. *Cinétique d'oxydation d'un composite carbone/carbone et influence sur le comportement mécanique*. PhD thesis n°1692, Université de Bordeaux I, 1997.

[6] D. Cho and B. I. Yoon. Microstructural interpretation of the effect of various matrices on the ablation properties of carbon-fiber-reinforced composites. *Composites science and technology*, 61:271–280, 2001.

[7] J. C. Han, X. D. He, and S. Y. Du. Oxidation and ablation of 3D carbon-carbon composite at up to 3000°C. *Carbon*, 33(4):473–478, 1995.

8 Y.-J. Lee and H. J. Joo. Investigation on ablation behavior of CFRC composites prepared at different pressure. *Composites: Part A*, 35:1285–1290, 2004.

9 J. D. Anderson. *Hypersonic and high temperature gas dynamics*. Mac Graw-Hill, 1989.

10 J. W. Metzger, M. J. Engel, and N. S. Diaconis. Oxidation and sublimation of graphite in simulated re-entry environments. *AIAA Journal*, 5(3):451–459, 1967.

11 T. A. Dolton, H. E. Goldstein, and R. E. Maurer. Thermodynamic performance of carbon in hyperthermal environments. In *AIAA 3rd Thermophysics conference*, Los Angeles, California, June 1968.

12 J. H. Lundel and R. R. Dickey. The ablation of graphitic materials in the sublimation regime. In AIAA Paper, editor, *AIAA 7th Thermophysics conference*, volume 72, San Antonio, Texas, April 1972.

13 J. A. Keenan and G. V. Candler. Simulation of ablation in earth atmospheric entry. In AIAA Paper, editor, *AIAA 28th Thermophysics conference*, volume 93, Orlando, FL, July 1993.

14 T. P. Roberts. Modelling gas/surface interaction processes of ablating wall boundaries associated with planetary entry. In *Proceedings of the second European symposium on aerothermodynamics for space vehicles*, pages 311–315, Noordwijk, The Nederlands, 21-25 November 1994.

15 M. A. Havstad and R. M. Ferencz. Comparison of surface chemical kinetic models for ablative reentry of graphite. *Journal of thermophysics and heat transfer*, 16(4):508–515, October-December 2002.

16 J. Nagle and R. F. Strickland-Constable. Oxidation of carbon between 1000-2000°C. In Pergamon Press, editor, *Proceedings of 5th conference on carbon*, pages 154–164, 1962.

17 F. P. Incropera and D. P. De Witt. *Fundamentals of Heat and Mass Transfer*. John Wiley and Sons, 5 th edition, 2002.

18 J.-F. Delesse. *Diffusion et réaction des gaz en milieux poreux*. PhD thesis n°2598, Université de Bordeaux I, 2002.

19 S. Torquato and I. Kim. Efficient simulation technique to compute effective properties of heterogeneous media. *Appl. Phys. Lett.*, 55:1847–1849, 1989.

20 A. Einstein. *Investigations on the theory of the brownian movement (Translated by A. D. Cowper)*. R. Fürth, Dover edition, 1956.

21 J. Sallès. *Caractérisation et colmatage des milieux poreux. Application au faciès calcaire du Dogger dans le Bassin Parisien*. PhD thesis n°604, Université P. et M. Curie, Paris 6, 1992.

THREE-DIMENSIONAL CHARACTERIZATION OF MICROSTRUCTURE AND INTERFACES

Dean J. Miller, Jon M. Hiller
Electron Microscopy Center and Materials Science Division
Argonne National Laboratory
Argonne, IL 60439

Edward L. Principe
Carl Zeiss SMT
Redwood City, CA 94065

ABSTRACT

Characterization of microstructure in three dimensions offers many advantages over analysis based on more traditional 2-D imaging, providing, for example, measurement of phase distribution, volume fractions, and interfacial area. A variety of approaches have been developed to provide this information, but the advent of focused ion beam (FIB) techniques, especially associated with dual ion- and electron-beam instruments (FIB-SEM), has greatly facilitated such 3-D characterization. In this work we have used FIB-SEM-based approaches to characterize the microstructure of high temperature superconducting tapes and fuel cell materials in three dimensions. In each case, phase distribution and interfaces play an important role in ultimate behavior, and 3-D reconstruction of microstructure provides a useful way to visualize these important microstructural characteristics. One key to such reconstruction is reliable, high quality volumetric data collection that can be interrogated to provide such 3-D information. The specific approach to data collection and analysis may vary depending on the microstructural features of interest. In this paper we outline these approaches using our work on high temperature superconducting tapes and fuel cell anode materials as examples.

INTRODUCTION

In materials science, the spatial correlation between microstructural features plays an important role in understanding physical behavior. In dispersion-strengthened materials, for example, the size and distribution of second phase particles is a critical parameter in mechanical strength. Likewise, in composites the phase fraction and distribution are both key to behavior. For example, in a composite of electrically insulating and conducting phases, electrical conductivity requires a continuous path through the conducting phase. While percolation models make useful predictions about such connectivity, it is desirable to be able to directly visualize such parameters, and 3-D reconstruction of microstructure offers this possibility.

3-D reconstruction based on tomography is a well-established technique, especially in the medical field where techniques such as x-ray Computer Axial Tomography (CAT) and Magnetic Resonance Imaging (MRI) are widely used for diagnosis. In materials science, many approaches for 3-D tomography have been developed. Each approach offers some advantages and poses some limitations. For example, x-ray tomography offers good depth of penetration and is non-destructive, making it an attractive technique for in situ studies. While x-ray tomography based on conventional "laboratory" x-ray source instrumentation is commercially available, that approach suffers from limited resolution, in the micron range. High-resolution x-ray tomography, with resolution on the order of hundreds of nanometers, has been developed, but is only possible using synchrotron sources.[1] At the other end of the spectrum, atom probe techniques provide exquisite structure and chemistry at the atomic scale, but with stringent requirements on specimen preparation and limitations on the volume of materials that can be explored, typically limited to less than 1/1000 of a cubic micron.[2] FIB-SEM based tomography provides information in between these length scales, with volumes exceeding 1000 μm^3 and nanometer-scale resolution possible (albeit not at the same time) as will be discussed in more detail further below.

FIB-SEM TOMOGRAPHY APPROACH AND CONCEPTS

The common feature of all tomographic methods is the acquisition of a series of 2-D images and their reorganization to reconstruct the 3-D object. The series of images can be obtained by imaging the object at different angles (tilt) or at different depths within the sample (series). Tilt-based tomography is typically utilized for x-ray based techniques. FIB-SEM based approaches utilize serial sectioning to obtain an image series, with the focused ion beam used to ion mill/polish into the sample while the scanning electron beam is used to image the surface yielding a sequence of images. The subsequent organization ("stacking") of those images yields a data set or data cube that can be interrogated in three dimensions. The serial tomography approach is illustrated schematically in Figure 1.

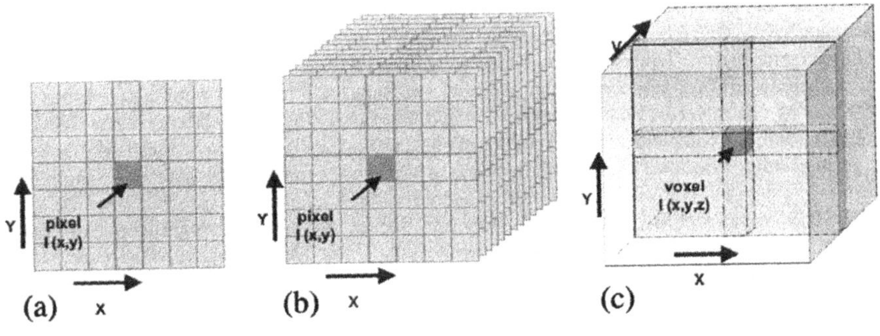

Figure 1: The serial tomography approach. (a) Each single image contains signal information, I, for each spatial coordinate pixel (x,y). (b) A series of aligned images form the basis for volumetric data. (c) Volume rendering "fills space" between sequential images based on the signal in each pixel. This volumetric unit is commonly referred to as a "voxel."

Serial tomography using sequential sectioning and polishing followed by imaging of the surface has been used in materials science for decades. Imaging can be carried out by many means, but SEM imaging is commonly used. Although the excitation volume in SEM can large, approaching a micron in range depending on the material and the incident energy of the beam, the escape depth for secondary electrons used for imaging is on the order of nanometers. Thus, resolution in the *x* and *y* dimensions has not been a major issue. However, mechanical approaches for sectioning yield large spacing between images in the v dimension (see Fig. 1), in the 10's of microns range at best, typically. The advent of FIB-SEM provides new levels of resolution based on the unique capability of the FIB to create closely spaced sections, separated by only nanometers, yielding the possibility of nanometer-scale resolution in all three dimensions.[3]

While nanometer-scale resolution is possible, FIB-SEM is extremely valuable over longer length scales as well. While there are limits on the volume of material that can be explored, dimensions of 10's of microns are easily accessible. This means FIB-SEM can be useful for materials with micron-scale grain sizes. There is no fundamental limitation to analysis over larger length scales within reason – it is primarily limited by instrument time. An important aspect of FIB-SEM is that conditions can be optimized such that the resolution is equal in all three dimensions.

A key step in the process is to acquire a series of images with know spacing along the v-axis (see Fig. 1). There are two fundamental approaches to this step: dynamic acquisition of images during milling, or sequential acquisition with interrupted milling. Conceptually, the two approaches are similar when one considers that milling of each plane surface is a stepwise process. With sequential imaging, a planar surface is milled and polished, then the ion beam is blanked while an image is collected. Following image acquisition, the ion beam is repositioned for the next milling step, and the process continues. Under dynamic acquisition, one acquires an image as the ion beam is milling the planar surface. Ideally, an image would be recorded at the end of the milling step such that the entire surface is polished to the same depth in v. In order to obtain high-quality images, this is accomplished by synchronizing the image acquisition to the milling process, initiating image acquisition after the region of interest is polished but while the ion beam continues to mill the same slice. Naturally, milling conditions adjusted appropriately to allow time for image capture before the next milling slice is initiated. In the work presented here, we collect a single frame over 5-10 seconds for each slice with a spacing v typically 5-10 nm.

While both processes are similar, we find some advantages to dynamic imaging. Dynamic imaging is less susceptible to errors in v since the ion beam is not blanked. Repositioning the ion beam after it is blanked is prone to larger errors than positioning the beam to a new location within the milling array due to inherent limitations of the ion beam scan generator. A second, more practical, advantage is that dynamic imaging offers improved image quality for samples that have insulating phases that charge under normal SEM imaging. The presence of the ion beam alleviates the charging effect and improves image quality.

For either approach, it is important to know the spacing between images, measured by v. This can be accomplished by milling fiducial marks on the sample that will remain visible throughout the milling process. An example of this approach is shown in Figure 2. In this case, lines are milled into the sample surface such that the spacing between the lines reflects the depth, v, into the sample. Under dynamic imaging, one can also use time as a measure of depth. This is rather useful since measurement of v for each image using fiducials can be time-consuming. Since the beam is not blanked under dynamic imaging, using time as a scale for v is reliable within the error of the scan generator of the FIB, typically sub-nanometer. We have been able to achieve agreement between these two approaches within 10nm over micron depths in v.

Figure 2 also reveals another important aspect of 3-D analysis, namely alignment of images (image registration). There is a horizontal displacement between Fig. 2a and 2b that is best noted by the center line in the series of fiducials. Less obvious, but also present, is a vertical displacement. Such displacements can arise due to specimen drift over the time frame of data collection, although a vertical displacement is assured due to the geometry of FIB-SEM as shown schematically in Fig. 2c. At the present time, image registration is most commonly carried out after data acquisition by software routines. For the work presented here, we used post-acquisition image registration based algorithms available in commercial software.[4,5] It is interesting to note that as instrumentation hardware and software improves, alignment of images *during* acquisition is becoming increasingly practical. Drift correction routines have long been available for microscopy applications such a energy dispersive mapping and spectral imaging. These routines use image recognition algorithms together with image shift hardware to re-center a region of interest that may have shifted due to drift. These routines have already been adapted for FIB-SEM to facilitate automated TEM specimen preparation, and now are being extended to image registration during tomography. Thus, image registration may be considered part of data acquisition or part of data processing, depending on the method used.

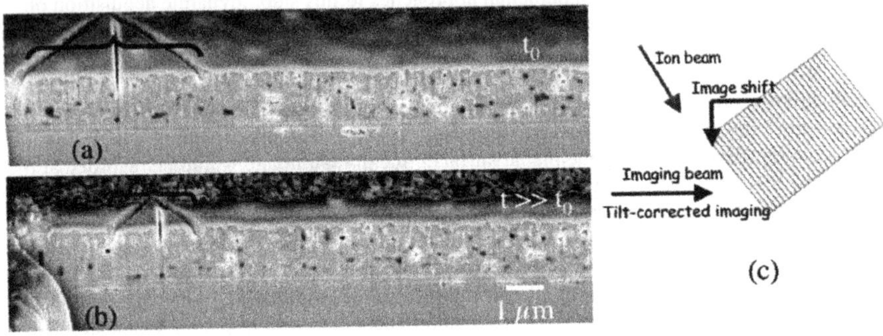

Figure 2: Measurement of section depth along the *v* axis using fiducial marks. Single lines are patterned into the sample surface before sectioning begins. Typically, a line parallel to the milling axis (v-axis) plus two lines ± 45° from that line are patterned. These lines are indicated by the braces in (a) which is designated t_0. After milling further into the sample along v as in (b), the separation in the milling plane between the fiducials is smaller and can be measured to determine separation between planes. (c) The geometry of FIB-SEM.

Once a data set consisting of a series of aligned images has been acquired, 3-D reconstruction can proceed, generally encompassing image processing, segmentation, and 3-D rendering. All of these steps can be carried out using commercial software, but image processing and segmentation typically require significant manual intervention even within automated routines. The goal of these steps is to delineate the microstructural features of interest. In this sense, it can influence data acquisition strongly as well. Consider the case of a two-phase material in which one would like to be able to differentiate the two phases readily. This would suggest adjusting image acquisition parameters for maximum contrast between the two phases, ideally binary. While this seems simple, in practice it can be difficult. As an example, take the case of a binary system of a single phase with pores. One could imagine the phase being bright while the pores are dark – 1's and 0's in a binary image. However, consider an image in which the pore has almost been cut through such that the solid phase is just below the surface. Since the underlying phase is also illuminated by the electron beam, it will exhibit a signal very close in intensity to the surrounding material, making such a differentiation difficult. There are many examples of similar situations, and moving to more complex microstructures, such as multiple phases, exacerbates the challenge. There are many references for image processing, including a comprehensive text.[6] In this work, we used processing routines available within the reconstruction software.

Figure 3 shows an example of the segmentation process using the program Amira. Figure 3a shows a raw image of a superconducting thin film cross-section consisting of a matrix phase, lighter second phases, and darker pores. In this example, the pores are of particular interest, and a threshold routine is used to delineate the pores in Fig. 3b. Comparing 3a and 3b, it can be seen that the automated routines do a good, but not perfect, job delineating the pores. For example, the arrowed pore in 3b is not fully captured by the automated routine. While the threshold criterion may be adjusted to improve the match on this image, one must recall that such a change may cause the matching in other images to become poorer due to changes in contrast or other factors. Fortunately, most of the available software packages offer manual tuning of individual features after threshholding. As an example, the boundary of the arrowed pore in 3b could be redrawn manually while the other defined pores remain unchanged. While this flexibility is important, it also illuminates the large degree of interpretation required in this process and the largest potential source of error. This challenge is not unique to 3-D reconstruction and has been considered carefully by the Microscopy Society of America, which has established a standard for image processing. Once the features of interest have been identified, software routines can segment the data as shown in Fig. 3c, in which all pixels outside of the delineated region of interest (pores) have been set to zero while all pixels inside the defined boundaries are unchanged. This routine is applied to all images in the data set. Further processing can be used to create a binary image if desired.

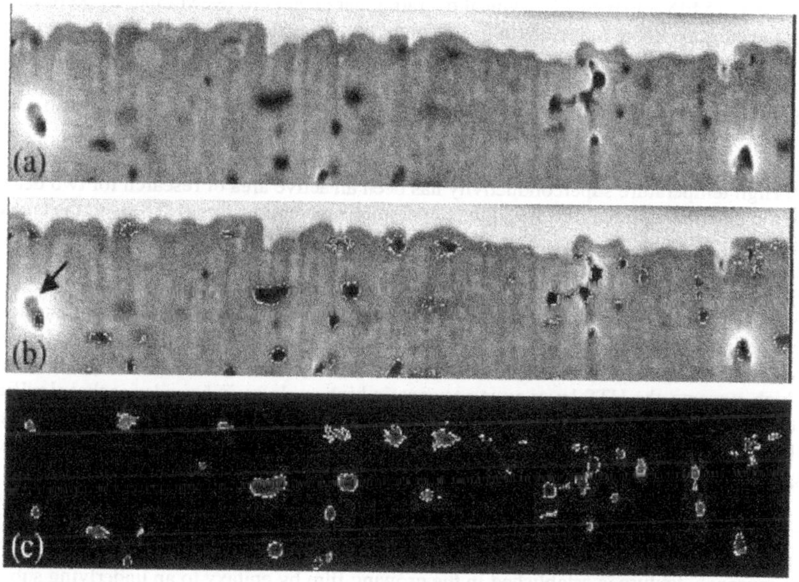

Figure 3: Delineation and segmentation of pores in a superconducting thin film. (a) Cross-sectional image showing the matrix phase, lighter second phases, and darker pores. (b) Pores delineated by an automated algorithm. (c) Image reduction, setting all pixels outside of the defined boundaries to zero. See text for details.

Once the data has been processed and segmented, many programs are available for 3-D reconstruction. Many algorithms are available to perform the reconstruction and the details of those are beyond the scope of this work, but have been the subject of other texts.[7] However, it is useful to comment on how the FIB-SEM approach impacts this issue. The 3-D reconstruction process basically involves interpolation, filling space between measured data points in each slice. Many algorithms have been developed to deal with the situation in which the resolution in v is much lower than that in x and y – the case where the spacing between "slices" is much larger than the resolution in each slice. In that situation, it is not obvious that a simple interpolation between slices is appropriate, and routines that consider changes in nearest neighbor pixels and even higher order correlations have been developed. In extreme cases, for example where the slice spacing is very large, it is obvious that this process can lead to errors on the order of image interpretation as described above. However, as noted earlier, FIB-SEM allows acquisition in which the resolution is equal in all three dimensions (or even better resolution in v compared to x and y, depending on acquisition parameters). Thus, FIB-SEM reduces the potential for this type of error.

THREE-DIMENSIONAL MICROSTRUCTURE OF HIGH-TEMPERATURE SUPERCONDUCTOR AND FUEL CELL MATERIALS

In the following section we will illustrate how the 3-D reconstruction process can provide valuable information using examples from high-temperature superconducting materials and fuel cell anode materials. In each of these systems, the relationship between microstructural constituents plays an important role in properties. In each case, data acquisition was carried out using a Zeiss 1540XB FIB-SEM. Typical operating parameters for SEM imaging were 10 keV with probe currents in the range of 100 pA. Tomographic sectioning by FIB was carried out using initial coarse milling (30 kV with high probe currents in the range of 1 nA) followed by polishing at successively lower probe currents to establish an initial surface of interest. Final milling for tomographic sectioning was carried out at lower probe currents, typically 50 pA.

3-D Microstructure in High Temperature Superconductors

High temperature superconductivity has been an active area of research for two decades. One of the potential applications for high temperature superconductors (HTS) is in the area of electrical transmission and storage, as well as in devices that protect the electrical grid from catastrophic failure, such as fault current limiters. A core technology for each of these applications is a conductor based on HTS materials, and "coated conductors" are the leading prospect for such HTS conductors.[8] A coated conductor, as illustrated in Figure 4, is a long-length tape incorporating an HTS film layer as the main current-carrying component. In coated conductors, the HTS layer is usually based on the $YBa_2Cu_3O_x$ compound (YBCO). In order to carry high currents, the HTS layer must not contain high angle grain boundaries that significantly limit current flow. In order to achieve the high degree of texture required, growth conditions must be carefully controlled and additional microstructural features, such as second phases or pores, must be controlled to avoid disruption of textured growth.

A common approach to produce the HTS layer in coated conductors is to use physical vapor deposition for YBCO deposition. Chemical vapor deposition using metal-organic precursors (MOCVD) is amenable to scale-up required for fabrication of long-length conductors. In this approach, texture is established in the growing film by epitaxy to an underlying substrate

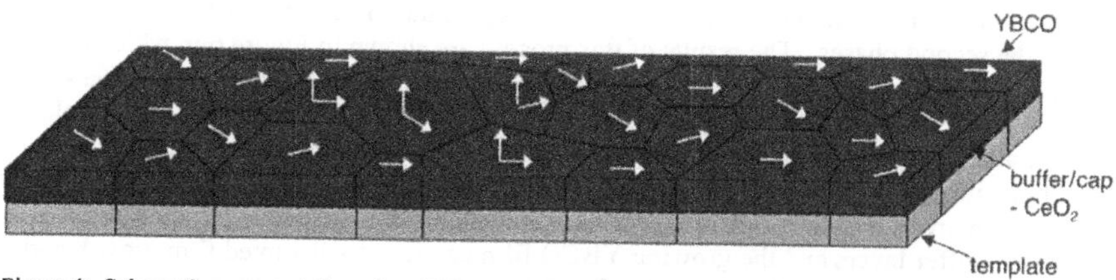

Figure 4: Schematic representation of an HTS coated-conductor. The YBCO film on the surface achieves a high degree of texture by growth onto a suitably textured template.

with a textured template layer. The epitaxial growth process can produce very high quality HTS films, as suggested by many studies on single-crystal substrates. However, second phase formation during growth can disrupt texture. The characterization of these films has been described in more detail elsewhere.[9] In this work, we use FIB-SEM to study the 3-D microstructure of an MOCVD-based film, with a focus on the role of second phases.

Figure 5a shows an SEM image of a cross-section of the HTS film. The microstructure consists of the YBCO matrix phase and various second phases. The presence of the second phases disrupts texture, so it is useful to know more about the distribution of the second phases. Figure 5b shows an image stack created from a series of such images acquired by FIB-SEM. The three-dimensional perspective is already clear in this figure. The data set was processed using software developed by E. Principe.[4] This processing included all aspects of 3-D

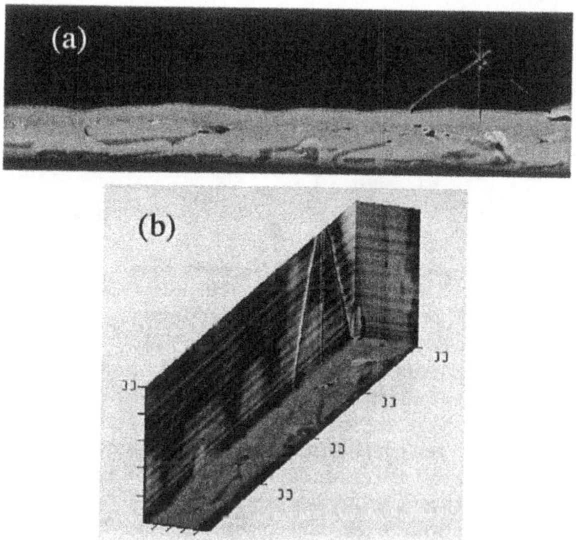

Figure 5: (a) Cross-sectional SEM image of an HTS thin film. (b) An image stack of the same film acquired by FIB-SEM.

reconstruction described above – image processing, segmentation, and reconstruction – in order to isolate the second phases. The results of this process are shown in Figure 6 in which the second phases are highlighted while all other phases are rendered transparent. Figure 6a shows the second phases from a general perspective. In Figure 6b the microstructure is viewed along an axis parallel to the substrate surface, with the substrate at the bottom. This perspective is useful in that it reveals only a few tiny intersections between the second phases and the substrate, a few of which are indicated by arrows. We have shown previously that an interaction between the substrate buffer layers and the growing YBCO film can lead to disrupted film growth and degraded superconducting performance.[9] The 3-D perspective offered here reveals that this takes place in local areas, consistent with those observations based on magneto-optical imaging.

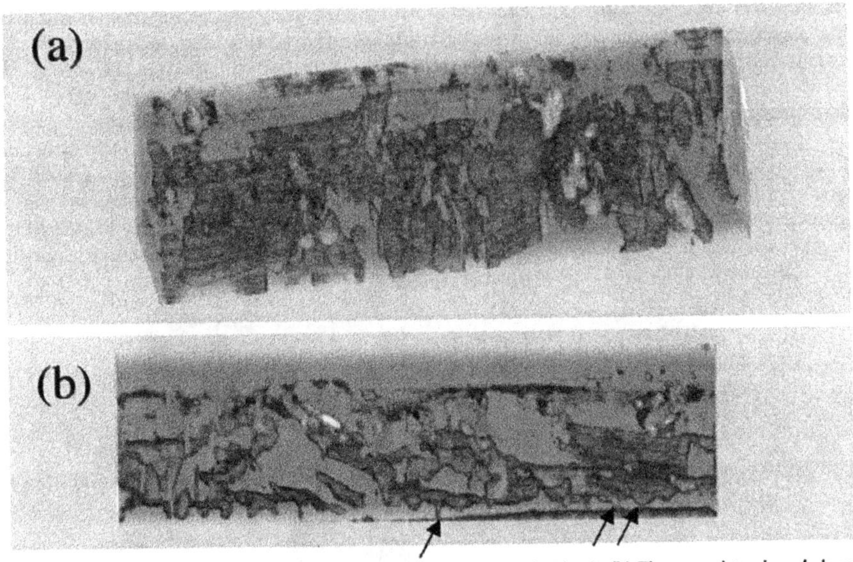

Figure 6: (a) 3-D rendering of second phases in the HTS film shown in Fig. 5. (b) The same data viewed along an axis parallel to the substrate (bottom). Intersections between the second phases and the substrate are arrowed.

3-D Microstructure in Fuel Cell Materials

Fuel cells that convert chemical energy into electrical energy are being studied intensely since they offer the possibility for higher energy conversion efficiency, yet with portability and lower environmental impact. A recent series of review articles provides a good description of many issues surrounding fuel cell materials and many references for further information.[10] In this work, we focused on the fuel cell anode, which must support simultaneous gaseous, ionic, and electronic conduction. Typcially, this is achieved in a single material by combining an electrically conducting phase with an ionically conducting phase and engineering to mirostructure to produce porosity that provides gaseous conduction. The most commonly used material is a composite based in nickel and yttria-stablized zirconia (YSZ). There are many microstructural aspects critical to performance, but in this work we focused on developing FIB-SEM to establish phase fractions.

In this experiment, we studied a Ni-YSZ anode composite. As noted in the introduction, percolation models can predict conductivity based on phase fraction. Therefore, it is useful to determine the relative phase fraction of each microstructural constituent, and 3-D reconstruction using FIB-SEM can yield this information. In this example, a data set was collected using FIB-SEM and processed using commercially available software Amira.[5] As in the case of the HTS materials, the reconstruction software was used for all aspects of image processing described earlier. The result of the reconstruction process is shown in Figures 7 a-c, which shows each constituent independently of the others. Based on this segmentation and rendering, the volume fraction of each component can be determined, in this case yielding 65% YSZ, 25% Ni, and 10% pore space. Ideally, pore space would be higher to promote more gaseous diffusion and to create more three-phase boundaries at which the reactions can take place, so this information is valuable input for processing. As noted earlier, 3-D reconstruction should make direct analysis of contiguity and connectivity for each phase possible, and that is the subject of further work.

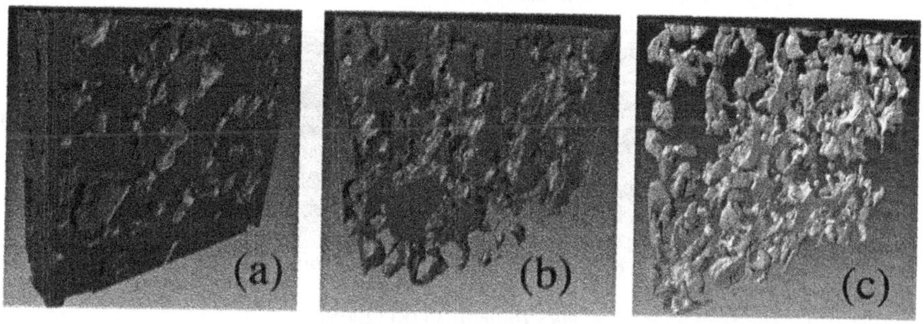

Figure 7: 3-D reconstruction of a Ni-YSZ composite showing segmentation of (a) YSZ, (b) Ni, and (c) pores.

CONCLUSIONS

We have used FIB-SEM-based approaches to characterize the microstructure of high temperature superconducting tapes and fuel cell materials in three dimensions. The key to such reconstruction is reliable, high quality volumetric data collection that can be interrogated to provide such 3-D information. In this paper we outlined the FIB-SEM approach to tomography and then demonstrated the approach with two examples. In high temperature superconducting tapes, second phases can form due to reactions with the underlying substrate materials, and 3-D reconstruction was used to show that the second phases formed adjacent to the substrate are found only in localized regions. In a fuel cell anode material, 3-D reconstruction was used to evaluate the relative phase fractions of each microstructural constituent. This information is important to refine the microstructure to improve properties.

ACKNOWLEDGEMENTS

This work was supported by the U. S. Department of Energy, Office of Science, under Contract W-31-109-ENG-38. Microscopy was carried out in the Electron Microscopy Center at Argonne National Laboratory.

REFERENCES

[1] Wang YX, De Carlo F, Mancini DC, McNulty I, Tieman B, Bresnahan J, Foster I, Insley J, Lane P, von Laszewski G, Kesselman C, Su MH, Thiebaux M, "A high-throughput x-ray microtomography system at the Advanced Photon Source," *Review of Scientific Instruments* 72 (4): 2062-2068 (2001).
[2] Miller MK, "The development of atom probe field-ion microscopy," *Materials Characterization* 44 (1-2): 11-27 (2000).
[3] see, for example, a collection of papers in "Focused ion beam (FIB) / dual beam applications in biological and physical sciences," *Microscopy and Microanalysis* 10 (2), pp 140 (2004).
[4] Carl Zeiss SMT software, E.L. Principe.
[5] Amira, Advanced 3D Visualization and Volume Modeling software, Mercury/TGS, www.amiravis.com.
[6] J.C. Russ, The Image Processing Handbook, CRC Press; 4th edition (2002).
[7] Natterer, F. The mathematics of computerized tomography, Society for Industrial and Applied Mathematics (2001).
[8] High Performance YBCO-Coated Superconductor Wires, *MRS Bulletin* 29 (8) (2004).
[9] Maroni, VA; Venkataraman, K; Miller, DJ; Trasobares, S; Lei, Y; Hiller, JM; Gray, KE; Vlasko-Vlasov, VK; Claus, H; Reeves, J; Rupich, MW; Zhang, W; Kodenkandath, T; Li, X, "Coordinated characterization of coated conductors," *IEEE Trans Appl. Supercon.* 15 (2): 2798-2802 (2005).
[10] Fuel Cells: The Next Evolution, *MRS Bulletin* 30 (8) (2005).

CHARACTERIZATION OF INTERFACES AND DEFECTS ON THE ATOMIC SCALE WITH THE CONJUCTION OF STEM Z-CONTRAST IMAGING AND EELS

L. F. Fu, N. D. Browning
Department of Chemical Engineering and Materials Science,
University of California Davis,
Davis, CA 95616
And
National Center for Electron Microscopy,
Lawrence Berkeley National Laboratory,
Berkeley, CA 94720

D. C. Kundaliya, S. X. Zhang, W. Ramadan, T. Venkatesan, S. B. Ogale
Center for Superconductivity Research,
2117 Physics Building, University of Maryland,
College Park, MA 20742

ABSTRACT

Interfaces between two dissimilar materials have long been known to dominate the performance of many heterointerface systems. Microstructural characterizations at the highest spatial resolution have always been a subject of great scientific interest to understand the physical properties of these systems. Commercially available high spatial resolution scanning transmission electron microscopes (STEM) have provided the possibility of widespread applications of electron microscopic characterization techniques for interface studies. For instance, the atomic and electronic structure of interfaces can be studied directly without computer simulation or reconstruction using Z-contrast imaging and EELS. In this paper, we discuss the practical aspects of the experimental STEM techniques and demonstrate the application of these techniques to the study of heterophase interfaces and defects in YBCO-$SrTiO_3$, Co:TiO_2-$LaAlO_3$, and Co:HfO_2-YSZ ceramic-oxides interface systems.

1. INTRODUCTION

The formation of interfaces between two crystalline or noncrystalline solids with different chemical, electronic, and structural properties exhibits both numerous challenges and opportunities for investigation of fundamental scientific issues [1-3]. Extensive heterointerface studies have been performed recently in for instance, superconductor-oxide interfaces, semiconductor-oxide interfaces, metal-oxide interfaces, and oxide-oxide interfaces [4-9] . These interfaces have a wide variety of applications, such as quantum well lasers, quantum tunneling devices, access memory devices, and high electron mobility transitors [10] . However, developement of these emerging technologies has been hindered by lack of knowledge on both the interface structure and resultant properties of these hetero-interfaces. For example, the interfacial defects formed between the film and the substrate may not present in three-dimensional forms, but their unknown properties can severely affect the physical behaviour of the designed devices or products [11]. Therefore, a detailed structural and chemical analysis is needed for understanding and improving the physical properties and quality control of the heterointerface.

In the past decades, a variety of microstructure characterization techniques have been applied to the structural and chemical analysis of heterointerfaces, including x-ray diffraction [12], scanning electron microscopy [13], transmission electron microscopy (TEM)[14], and scanning transmission electron microscopy (STEM) [15]. Among them, STEM is well known to possess the ability to form a Z-contrast image, in which intensity is proportional to the square of the mean atomic number of the probed volume and is directly interpretable [16]. This ability of STEM distinguishes itself from others and makes it a very powerful instrument for microanalysis at high-spatial resolution. Moreover, the instrumental and technique development in the STEM field has fostered the conjunction of electron energy loss spectroscopy (EELS) and high angle annular dark field (HAADF) Z-contrast imaging and their widespread use for nearly atomic level chemical analysis at interfaces and isolated defects [17-20].

In this paper, we will introduce these novel imaging and analytical techniques with their capabilities and the requirements for the instruments to obtain the atomic scale information. Particularly, we concentrate on the methodology behind the use of EELS and Z-contrast imaging in combination in the STEM. Then, we will demonstrate that a wide range of materials and interfaces, including superconductor-oxide interfaces, diluted magnetic semiconductor-oxide interfaces, and ceramic oxide-oxide interfaces, can be studied by these techniques.

2. EXPERIMENTAL TECHNIQUES

The experimental results demonstrated here were all obtained using a 200kV Schottky field-emission gun (FEG) monochromated FEI Tecnai F20UT STEM/TEM at the National Center for Electron Microscopy (NCEM) at Lawrence Berkeley National Laboratory (LBNL). This microscope is equipped with an ultra twin objective lens pole piece, which enables complete materials characterization by combining high resolution TEM with high resolution STEM and advanced spectrum imaging in EELS. The control system of the microscope is very different from that of traditional microscopes, and a computer is the main control and analysis interface. The integrated user-friendly interface allows the user to customize the control panels and define the desired digital acquisition configurations. This microscope also features a 2048x2048 CCD camera positioned after the Gatan Imaging Filter (GIF) system that can be used for both dedicated energy loss spectroscopic analysis and energy filtered imaging. The Schottky field emission source in this microscope has a much larger emission area than cold field emitter conventionally used in dedicated STEM [21]. It combines the high brightness and low energy spread of the cold field emitter with the high stability and low beam noise of the thermal emitters. The emitter in this microscope gives a low energy spread (< 0.7 eV at useable extraction voltages) and very small probe sizes (< 0.2 nm), which enables an ultimate high spatial resolution of 0.14 nm in STEM mode with the equipped ultra-high resolution STEM HAADF detector. In this section, we will briefly introduce the main aspects of STEM imaging and spectroscopic analysis used in our study. Their experimental operation details will be made clear in the following sub-sections.

2.1 STEM Z-contrast imaging

Atomic resolution STEM HAADF Z-contrast imaging is a new characterization technique developed in recent years and has been applied to characterize the structure of interface, grain boundaries, and different defects [22-24]. The electron beam runs in reverse between STEM microscope and conventional transmission electron microscope (CTEM) (Figure 1). The electron ray runs downwards and upwards in STEM and CTEM, respectively. Compared to conventional optical microscopy or CTEM, image formation in STEM is based on a different principle. In STEM, the electron beam is focused by the condenser system to a small probe, and then scanned in a rectangular pattern across the specimen. The interaction of the electron beam with the specimen gives rise to various signals, such as transmitted electron, back scattered electron, secondary electron, X-rays, etc. The illumination system of a STEM consists of gun lens, condenser lens, and the up pole-piece of objective lens. The function of the illumination system is to de-magnify the source to obtain a bright and small spot (0.2nm or better) on the specimen, which also determines the ultimate spatial resolution. Different from a TEM column, the lenses of electron optical system in STEM include a scan generator to scan the probe across the specimen and to drive the display on monitor. The magnification of the scanning mode in a STEM system is the ratio between the final image display (horizontal length D) and the area scanned on the specimen (horizontal linescan-length d).

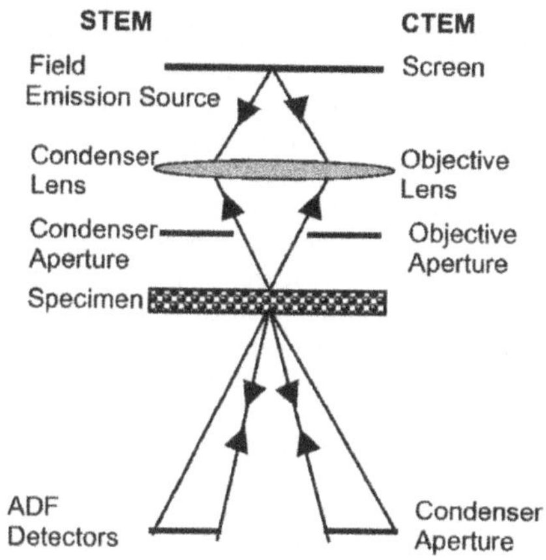

Figure 1 Ray diagram of STEM and CTEM. The electron ray illuminates downward and upwards in STEM and CTEM, respectively.

Z-contrast images are formed by collecting the high-angle scattering (75-150 mrad at 100kV, 40-100 mrad at 200kV) electrons on a high angle annular detector below the specimen (Figure 2). To achieve the highest resolution in STEM Z-contrast imaging, the tuning and alignment of the electron beam is the key. This optimization of the electron probe can be obtained with the means of the electron "Ronchigram" or "shadow image"[21]. A Ronchigram is formed by collecting the transmitted electrons as a function of angle on a charge-coupled device (CCD) or equivalent device and can be thought of as a shadow image of the sample [25]. For an amorphous sample, this resembles a distorted bright field image of the sample and directly reflects the symmetry of the microscope aberrations. Axial astigmatism can be very accurately corrected by exciting the stigmator coils so that these Ronchigram features are circularly symmetric. The coma free axis can be clearly defined by this position and all alignment of detectors and apertures can be performed with respect to this position. If there is a misalignment of the beam between condenser and objective lenses, there will be a periodic translation of Ronchigram features, such as the wobbling of the images. As a result, the probe can be aligned most readily to the coma-free axis. Once the electron beam is aligned along the optic axis in the illumination system, the specimen is on the condenser lens focal plane, an optimum focused probe can be formed on the surface of the specimen. Then, this convergent incident probe is scanned across the region of interest, and if the specimen is thin enough to be electron transparent, the scattered electrons can be collected as a function of the probe position. The annular detector below the specimen (i.e., large inner angle, large detector area) acts as a high-pass or Bloch wave filter, which means that only the electrons sharply peaked around the illuminated position can contribute to the image[21]. With these settings the Z-contrast image is essentially incoherent, i.e. a direct image of the structure is acquired. As the scattering amplitude at high angle is essentially Rutherford scattering, the measured intensity is proportional to the square of average atomic number of the illuminated atomic column ($\sim Z^2$). Once the sample is oriented with one of its major zone axes parallel to the incoming electron probe and the interatomic spacing is larger than the full width at half maximum(FWHM) of the beam, the atomic columns are illuminated individually and atomic resolution can be achieved [26].

A distinguished advantage of STEM Z-contrast imaging is that the phase problem associated with the interpretation of CTEM images is eliminated. In STEM, detecting the scattered intensity at the high angles and integrating over a large angular range effectively averages coherence effects between neighboring atomic columns in the specimen. Thermal vibrations reduce the coherence between near neighbors, a second order effect. This allows each atom to be considered as an independent scatterer. For very thin specimens where there is minimal dynamical diffraction, the detected intensity consists of a convolution of the object function with the probe intensity profile and is proportional to the square of average atomic number of the illuminated atomic column. This result also holds true for thicker specimens since it has previously been discussed that STEM probe forms narrow spikes around the atomic column as it propagates [24]. Therefore, the fact that changes in the sample thickness can be observed as an increase in the background intensity rather than a contrast reversal as in CTEM. The intensity of the Z-contrast images is mainly related to chemical compositions and is therefore more readily directly interpretable than HR-TEM micrographs. The atomic structure in the regions of the complicated interfaces, grain boundaries, and defect structures can be determined directly from the images, without time-consuming image simulation.

2.2 Electron energy loss spectroscopy (EELS)

As shown in Figure 2, the annular detector only collects the high angle scattered electrons to form Z-contrast images; however, the low angle scattered electrons can still pass through the hollow area of the annular detector. The electron-optical setup of this microscope allows us to use the low-angle scattered electrons that do not contribute to the Z-contrast image for EELS. Since these two techniques do not interfere, this means that the Z-contrast image can be used to position the focused electron probe at the desired spot or particular structural features on the sample for acquisition of an EEL spectrum.

To analyze the interfaces, grain boundaries and defects, it is needed to correlate the spectrum directly with the structural features seen in the image. The correlation of the Z-contrast image and the EEL spectrum allows the image to be used as a structural model to interpret the fine-structure of the spectrum. To achieve this, it is essential that the spectrum and the image have the same spatial resolution. For crystalline materials, zone-axis orientations, provided that a large collection angle is maintained, coherent effects will be averaged and, like the Z-contrast image, the description of the spectrum in terms of a convolution of the probe with an object function is still valid [27]. The probe channeling effect, discussed for Z-contrast imaging, still plays an important role in atomic resolution EELS as the electron probe is focused and scattered along the atomic columns [23, 28]. Therefore, the spectra can be acquired from each atomic column separately if the probe is smaller than the atomic spacing. The spatial resolution of spectra is defined by the probe size on the surface of the specimen. This means the spatial resolution of the spectrum can achieve the atomic level even with the large collection apertures. The only limiting factor in this resolution is the specimen drift due to stage instabilities in the microscope. With the fully computer-controlled, and high stability CompuStage in the Tecnai, we are able to obtain a drift less than 0.5 nm/minute. For the acquisition times of less than 5 s per spectrum in all practical applications discussed here the broadening effect is not significant. Using the microscope lens setup for high resolution with enough beam-current (>100 nA), we are able to routinely achieve an energy resolution of 0.5 eV using the FHWM at the zero-loss peak at a dispersion of 0.05eV/pixel for the EEL spectra. By acquiring images before and after each EEL spectrum acquisition, the presence of specimen drift with respect to the electron probe can be checked. Therefore, we can take several spectra at similar positions to maintain an interpretable signal-to-noise ratio.

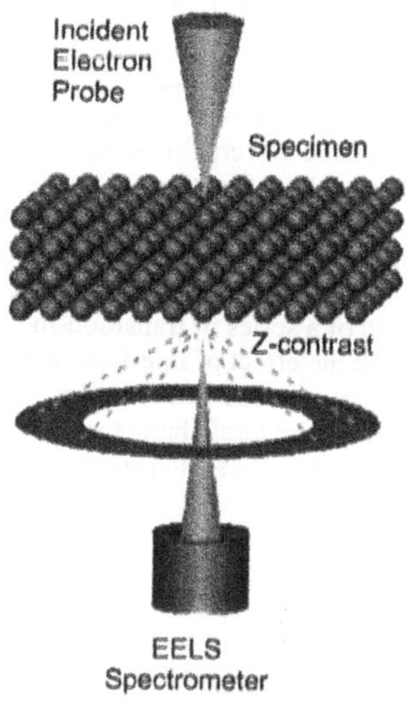

Figure 2 The schematic diagram of the post-specimen electron optics to explain the principle of STEM Z contrast imaging combined with EELS.

The physical principle behind EELS relates to the interaction of the fast electrons with the specimen, which leads to either collective excitations of electrons in the conduction bands (plasmons), or discrete transitions between atomic energy levels[19]. For the materials we are studying here, we concentrate on the highly localized discrete atomic transitions, which allow both compositional analysis (elemental distributions) and the degree of bonding hybridization (information on local electronic structure and valence change). For the interface and surface, we are primarily interested in the bonding between oxygen and the metallic atoms and the chemical segregation of heavy elements. To investigate this, the near edge fine structure of the oxygen K-edge is very important since it contains valuable information about the nearest-neighbor bonding, and the hybridization of O 2p with metallic sp band. The metallic transitions (Fe, Co, Ni, Ti, V, and Cr) from $2p^63d^m$ to $2p^53d^{m+1}$ or $3p^64d^m$ to $3p^54d^{m+1}$ reflects the number of available orbitals for the hybridization with the oxygen bands. All these energy regions of interband transitions lie in the sensitive part of EEL spectrum; therefore, we can correlate directly the bonding arrangement and metal valence with atomic structure observed in the Z-contrast images.

3. APPLICATIONS

Transition-metal oxides are well known to exhibit various intriguing physical properties, such as colossal magnetoresistivity and superconductivity, ferroelectricity, and ferromagnetism [3, 8-10]. However, the properties of these oxides are highly sensitive to changes in the external electric and magnetic fields, light, and carrier concentration. Through the fabrication of heterointerface structures composed of the transition-metal oxides, we are able to control the physical properties and therefore, develop new functional electronic devices. In fact, there have been various heterojunctions fabricated using perovskite oxides or similar structured oxides. Their exciting properties have been commercially applied to the application of new functional devices, such as magnetic storage or sensors, nonvolatile computer memory, and magnetoelectronic devices. The ultimate drivers for research in these devices-that devices be "usable, durable, and reliable"-ultimately define the technological and scientific challenges[6].

To achieve these goals, the growth and microstructural characterization of the heterointerface and the measurement of the physical properties of the interfaces are of primary importance. In this work, we investigate the interface and surface quality as well as the distribution of carrier concentration of a series of transition-metal oxide systems using STEM and EELS. Our aim is to understand the effect of interface or surface segregation on the defect chemistry in the oxide heterointerface systems as a function of different growth conditions. Through the correlation of their microstructural characterization with their physical or electric properties, we hope to improve their properties and prompt their practical applications in electronic devices.

3.1 YBCO-$SrTiO_3$

YBCO, chemical formula $YBa_2Cu_3O_{7-\delta}$, is a high-temperature superconductor(HTS) material with a superconducting temperature of 94K[29]. It was the first superconductor to break the liquid nitrogen barrier, i.e. the first to be superconducting at a temperature high enough to permit the use of liquid nitrogen as a coolant. The discovery of HTS in this compound has led to a large number of experimental as well as theoretical studies related to extraordinary properties of this material[2, 30]. YBCO has a so-called perovskite structure which refers to a kind of layering usually observed in this complex metal oxide. The lattice is composed of alternate perovskite layers ($ACuO_3$) where A is a rare-earth or alkaline-earth element (e.g., Y or Ba in YBCO).The dimensions of a single unit cell of YBCO are a = 3.82 Å, b = 3.89 Å, and c = 11.68 Å. The term 7-δ in the chemical formula implies a slight deficiency of oxygen. If δ=0, the lattice is in the orthorhombic phase whereas in the case of δ=1, the material has a tetragonal structure. The metal Cu ions in the YBCO compounds are usually partially oxidized and comprised of Cu^{2+} and Cu^{3+} ions, depending on δ, which leads to holes of positive charge within the lattice [5]. Thus, YBCO is also a p-type superconductor. Besides this, its electronic structure can be controlled by adjusting the doped carrier concentration in the conduction layers. Therefore, it should be very interesting to study the growth and physical properties of p-n junctions comprised of YBCO with different carrier concentrations. To ensure their growth in high quality, atomically flat, the characterization of the interface in the heterostructures with the means of STEM imaging and EELS is of primary importance. At the same, the microstructural characterization of such interfaces is also meaningful to understand the formation and properties of the p-n junctions and thereafter prompt their widespread applicability in microelectronics technology.

Since we are mainly concerned with p-n junctions with a heterointerface of transition-metal oxides, electron-doped $SrTiO_3$ is selected for n-type conductor. Although $SrTiO_3$ is intrinsically a wide band gap insulator with an indirect band gap of 3.2eV, doping a small amount of elements with extra electrons(e.g., Nb, La, or In) into this material makes it conductive and superconductive at low temperature [16, 29]. Another interesting character of this material is its photo-active property [20]. Large photocarriers and photocurrent have been observed in the transition-metal-doped $SrTiO_3$ when it is illuminated with ultraviolet light. In addition, the broad band luminescence around the visible emission region is also reported in this material [10]. Therefore, transition-metal-doped $SrTiO_3$ is a good candidate material for the study on the p-n junction interface acting as an n-type semiconductor.

In this study, commercially procured high quality single crystalline (001) Nb doped $SrTiO_3$ single crystal substrates (Nb concentrations of 0.5 wt%) were used for film growth. The YBCO films were grown by pulsed laser deposition (PLD) technique. A pulsed KrF Excimer laser was used for ablation. The films of optimally doped YBCO were grown at 200 mTorr oxygen pressure at the substrate temperature of 800 ^{0}C and then cooled to room temperature in oxygen pressure of 400 Torr. This growth condition is known to yield films with good crystallinity and superconducting properties. Then, the growth quality of the interfaces was further investigated by high resolution STEM using the cross-section of the YBCO thin film samples.

Figure 3a shows a typical low magnification cross-sectional STEM Z-contrast image of YBCO thin film sample. The YBCO layer has a brighter contrast than the Nb doped $SrTiO_3$ substrate due to its higher average atomic number. The interface between YBCO thin film and Nb doped $SrTiO_3$ substrate is clearly seen to be abrupt and epitaxial. The YBCO thin film layer has a thickness of around 90nm, which is consistent with the thickness predicted with the growth condition. A large amount of planar defects, appearing as dark lines in Z-contrast images, are present in the YBCO thin film which indicate that the average atomic number is reduced at these defects. The high resolution Z-contrast image in Figure 3b shows the structure of these defects in detail. These defects are alternately grown in the gap of two or three unit cells along a [001] orientation and appear parallel to the interface. These planar defects have been defined before as a type of stacking faults, which are associated with the layered structure of YBCO [31]. Different from the "perovskite block" consisting of the sequence "BaO-CuO_2-Y-CuO_2-BaO", two Cu-O chains rather than one are grown resulting into a local transition from the $YBa_2Cu_3O_7$(123) structure into $YBa_2Cu_4O_8$(248) structure. The two Cu-O chains are related by a ½[100] glide symmetry provided that the view orientation of YBCO film in Figure 3b is [010]. In the Z-contrast intensities, Ba columns show the highest intensity followed by Y and Cu columns. A closer inspection of the dark lines in the Z-contrast image with a higher resolution also confirms that the dark lines contain an extra Cu-O plane in this view orientation.

In order to investigate the electronic structural and compositional changes, EELS analysis has been performed. The distribution of Nb in the $SrTiO_3$ and at the interface was studied by probing the film via tens of EELS line scans across the interface starting with recording two signals on the YBCO film. A typical series of EEL spectra and the regions which the spectra acquire from are shown in Figure 4a and 4b. The range of energy for the line scan EELS analysis starts at 150 eV and ends at 535 eV in order include the signals of oxygen, niobium, and titanium for comparison. Proceeding from direction indicated by the arrow in Figure 4a, the first signal of EELS has been recorded on the YBCO and includes the O K-edge at 532 eV, the Y $M_{4,5}$-edges at 175 eV, Y $M_{2,3}$-edges at 300 eV and 312 eV. The oxygen signal for the following spectra from YBCO side to $SrTiO_3$ side significantly changes owing to the change in composition and nearest atomic coordination. At the same time, the Y $M_{4,5}$-edge and $M_{2,3}$-edges disappears while the Sr $M_{2,3}$-edges at 269eV and 280 eV appear consistently in the substrate side. The broad and steep features starting from 205 eV to 235 eV correspond to the edges for the Nb-$M_{4,5}$ transitions. The consistently same features from the first point near interface to the farmost point imply a uniform Nb distribution in the $SrTiO_3$ substrate. High-resolution Z-contrast imaging throughout the film confirms that there is not a segregation layer of Nb presented at the interface.

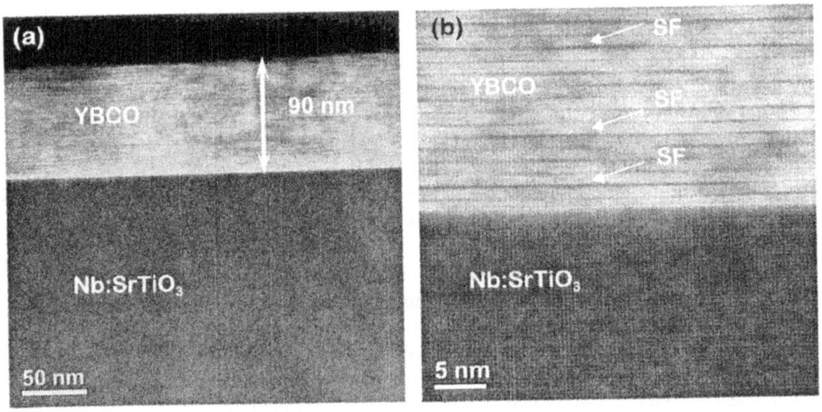

Figure 3 (a) A cross-sectional STEM Z-contrast image at large length scales of the YBCO thin film grown on Nb doped $SrTiO_3$ (001) substrate. (b) A high resolution STEM Z-contrast image shows the epitaxially grown interface and the existence of stacking faults in YBCO thin films.

As for the large variety and density of stacking faults found in the YBCO thin films, they are thought to come from a consequence of the non-equilibrium growth involved in the deposition process[31]. When the pulsed laser treats the reactive components during the growth process, the species are frozen into a relatively metastable configuration, where the diffusion and atomic arrangement are limited in short range. Therefore, the basic perovskite layering is disturbed and the structural changes take place mainly in the layers between two or three units.

The formation of these extra Cu-O chains affects the oxygen contents locally and might result into the doping carrier pinning center in the YBCO system. As a result, the superconducting properties are possibly changed by the formation and distribution of these stacking faults, similar to that of the 248 YBCO structures. The presence of these stacking faults gives us a promising prospect that the nonequilibrium laser deposition process can be used to control the metastable structure with novel superconducting properties. The EELS analysis across the interface found that the interfaces are chemically sharp within subnanometer scale, which confirmed the formation of high quality p-n junction heterointerface. The Nb distribution in the $SrTiO_3$ substrate was found to be quite uniform and nearly unchanged at the interface. This result implies that no a thin Nb depleted layer seems to be formed at the interface by the Nb diffusion during film growth. The good rectifying current-voltage characteristics of such epitaxial junctions reported previously can be intrinsically described from the Schottky junctions between YBCO and Nb-doped $SrTiO_3$ substrate[32].

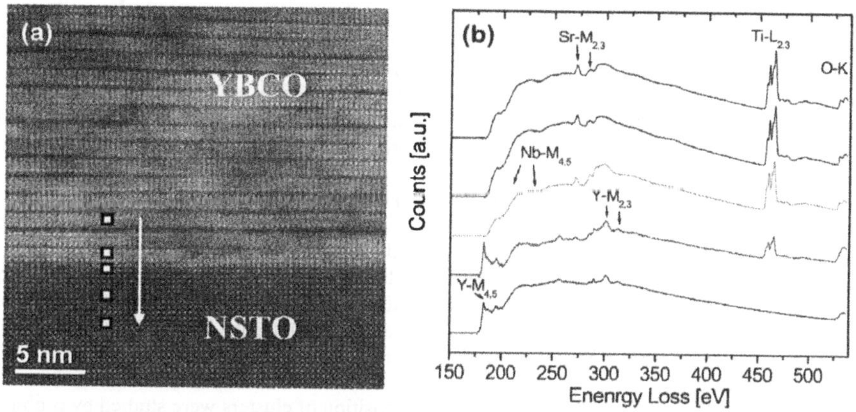

Figure 4 (a) A high resolution cross-sectional STEM Z-contrast image showing the high spatial resolution EELS acquisition area and acquisition direction. (b) EEL spectra acquired across the interface in YBCO thin film samples showing the distribution of Nb element and others.

3.2 $Co:TiO_2$- $LaAlO_3$

In order to explore new electronics in the booming semiconductor industries, it is interesting to combine both features of electrons: the charges and the spins. This exploration results into the naissance of spintronics (or spin electronics) which mainly exploit spin properties instead of or in addition to charge degrees of freedom. Based on this, a variety of semiconductor materials, called diluted magnetic semiconductors (DMS) have aroused much attention due to their application potential in the rapidly developing area of spintronics[33-35]. The popularity of DMS materials lies into their interesting characteristics: semiconducting and magnetic. Such a compound can be thought as an alloy between a non-magnetic semiconductor and a magnetic element. Recently, the occurrence of room temperature ferromagnetism (FM) using a combinatorial thin film library of TiO_2 thin film doped with various 3d transition metals has been reported by Matsumoto et al[36]. This work has attracted great interest in the oxide-DMS community and since that, many Co-doped TiO_2 thin films grown by various deposition methods have been reported with various properties. However, the fundamental issues in the DMS materials, such as the origin and the nature of FM, are still in controversy. Although carrier-induced interaction between the magnetic ions can partially explain the origin of the FM, this explanation is not extensively accepted [33]. Beside this, the magnetic properties of Co-doped TiO_2 films are found to depend critically on the Co distribution. At the same, the distribution of Co inside the TiO_2 is also strongly dependent on the growth process [37]. In this regard, the structural qualities of the diluted magnetic semiconductors play an important role in maintaining the spin character of the electron and its magnetic properties in the devices. To fully develop the DMS materials for spintronics applications, it is therefore necessary to characterize and understand the properties of the DMS thin films as a function of the growth conditions.

Thin films of anatase $Ti_{1-x}Co_xO_2$ (x= 0.03) studied in this work were grown on (001) $LaAlO_3$ (LAO) single crystalline substrates by a pulsed laser deposition method with an excimer laser (KrF, wavelength=248 nm) at 875 °C at an oxygen partial pressures of $1x10^{-5}$ torr. The laser energy density was ~ 2 J/pulse and the repetition rate was 10 Hz. The growth rate was controlled to be low to ensure the Co homogeneity. After the growth, the structures of the films were characterized by STEM Z-contract imaging and EELS using Tecnai at 200 kV. Core–loss EEL spectra were recorded with a dispersion of 0.2 eV/channel to contain more energy windows and an exposure time of 8s. The background for each core-loss spectrum was subtracted by a power-law fitting method.

Figure 5a shows a typical micrograph for the $Ti_{0.97}Co_{0.03}O_2$ sample. The micrograph is deliberately shown on a coarser length scale to check the presence or absence of any distributed clusters. The results revealed the formation of nanoclusters inside the film. The diameter of clusters varied from 10nm to 60nm. The high resolution STEM Z-contrast micrograph in Figure 5b shows the structural details of nanoclusters near the interface. In the micrograph, the atomic structure of nanoclusters is shown to be similar to that of TiO_2 matrix, except the Z-contrast change between them. This means that the clusters are composed of heavier elements than Ti in anatase-like structure.

The distribution of Co in the film and the composition of clusters were studied by probing the film via tens of EELS scans across the sample starting from the interface and subsequently probing the film in 10 nm steps in direction of the film surface. Some points were selected to stop on the nanoclusters. A typical such series of EELS scans is presented in Figure 6b and the corresponding acquisition region is demonstrated in Figure 6a. The range of energy for the line scan experiments starts at 520 eV and ends at 855 eV in order to include the signals of oxygen, cobalt, and lanthanum. The first signal has been recorded on the interface and includes the O K-edge at 532 eV and a weak La $M_{4,5}$ edge at 832 eV and 849 eV. There are no obvious Co signals presented in the film. Remarkably, the cobalt signal was found only on the nanoclusters in the sample. The white lines intensity ratio for the Co $L_{2,3}$-edge have been observed at 795 eV and 780 eV respectively implying an oxidized state of Co at 2.3+. Notably the intensity of both Ti $L_{2,3}$-edges and O K-edges on the nanoclusters is weaker relative to that on the pure TiO_2 film. Moreover, the fine structure of oxygen K-edge on the clusters is quite different than the signal in the TiO_2 film, implying a chemical bonding change as well as the formation of oxygen vacancies in the clusters. Considering all these factors, we deduce that those clusters might be Co-rich titanium oxide and exclude that they are metallic Co clusters from the valence state as well as the atomic structure shown in Figure 5b.

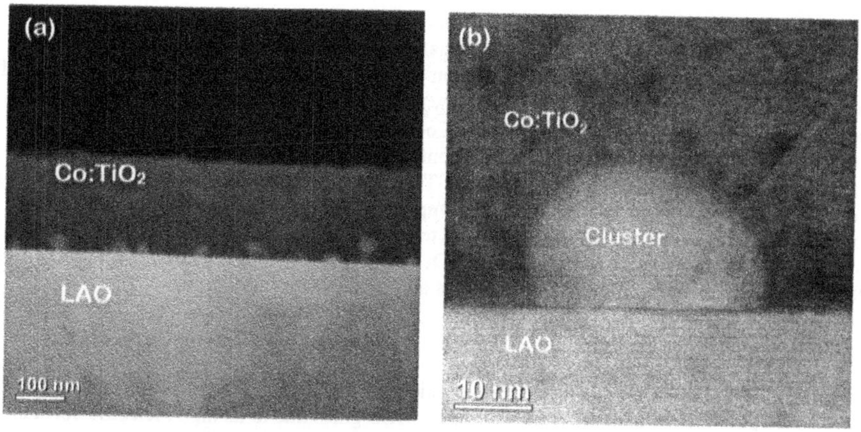

Figure 5 (a) A low magnification cross-sectional STEM Z-contrast image showing the existence of nanoclusters at the interface. (b) A high resolution Z-contrast image of the nano-sized clusters shows that they have a nearly epitaxial growth relationship with TiO_2 film.

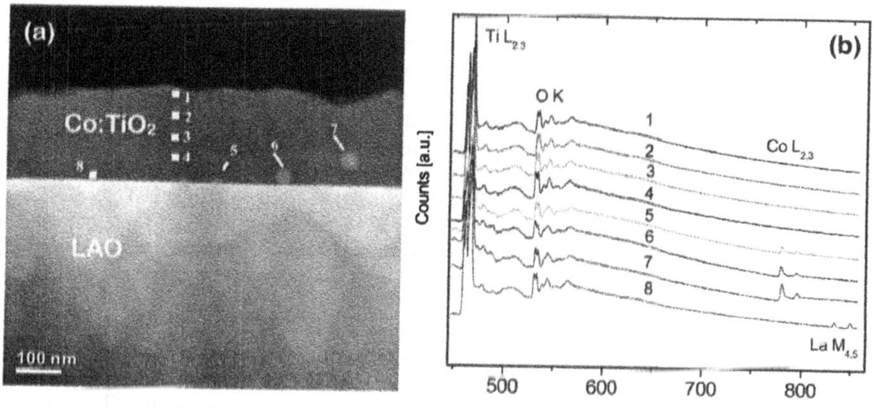

Figure 6 (a) A cross-sectional STEM Z-contrast image showing the high spatial resolution EELS acquisition area in the Co:TiO_2 thin film. (b) EEL spectra acquired across the interface in Co:TiO_2 thin film samples showing the distribution of Co element and others.

From the Z-contrast micrographs we obtained, it was shown that the density of the nanoclusters was much higher at the interface than in the film. This implies that the interface can work as preferential nucleation sites for clusters due to its instability. The formation of Co clusters should be the main origins for the observed FM in this film, which can be called extrinsic origins for FM [33]. It was suggested that the growth conditions, particular the oxygen pressure, play an important role in the formation of Co clusters and thus lead to a wide range of magnetic moments 0.3-1.7 μ_B/Co. The oxygen vacancies in the anatase TiO_2 film grown at low oxygen pressure will help the diffusion of the Co ions, resulting in the formation of the nanoclusters. In our case, oxygen pressure is controlled to be $1x10^{-5}$ torr, which is not quite low, compared to other deposition process reported. Thus, other factors resulting into the formation of Co clusters under growth conditions are not completely ruled out. For example, under certain conditions, highly Co-doped TiO_2 anatase tend to form clusters nucleated on the epitaxial interface of anatase TiO_2 grown on (100) $SrTiO_3$ due to limited solid solubility [37]. Further studies are required to clarify these possibilities and improve this film growth.

3.3. Co:HfO_2-YSZ

HfO_2 (Hafnia) is a promising insulating oxide that can be considered a replacement for SiO_2 as the gate dielectric in the next generation MOS gates due to its high dielectric constant and good thermal stability with silicon [38]. On the other hand, HfO_2 is also an attractive oxide which may enable the integration of CMOS with spintronics technology when introduced a magnetic response[39]. Recently reported intriguing ferromagnetism (FM) phenomenon in pure HfO_2 thin film has made it getting closer to this goal and aroused much attention to the study of its physical properties and mechanisms[40-42]. Generally, ferromagnetic insulators are rare since the spins tend to couple with each other antiferromagnetically in insulating compounds. However, it is feasible to introduce room temperature FM by doping transition metals into some semiconducting oxides. For example, the Fe, Co, and V doping in the ZnO matrix and Co in TiO_2 matrix have been reported to exhibit a room temperature ferromagnetic properties [43,44]. In this work, we are interested in the investigation of doping Co in HfO_2 in the thin film form to verify if such a doping can result in some kind of insulating ferromagnets.

High Purity HfO_2 (Cerac 99.99%) and CoO (Aldrich 99.99%) powders were mixed thoroughly and the mixture was heated at 1100 °C for 18h followed by grinding. The resulting powder was compacted in to a disc and sintered at 1175 °C for 20h. $Hf_{0.95}Co_{0.05}O_2$ target was used for PLD thin film growth on <100> Yttrium stabilized ZrO_2 (YSZ) substrates at 850 °C using a KrF excimer pulsed laser (wavelength = 193 nm). Growth oxygen partial pressure during growth process was 1×10^{-6} Torr.

A STEM Z-contrast micrograph of the cross-sectional sample is shown in Figure 6a. The image shows an epitaxially grown interface of the film with no formation of clusters. The thickness of HfO_2 thin film is determined to be 40nm, which is in consistent with the RBS-channeling spectroscopy experiment. High resolution imaging analysis across the film shows that both the film and interface are free of clusters or interfacial defects. Line scanning EELS analysis was employed to probe the atomic structure and bonding characteristics through the whole film. Tens of EELS line scans were performed across the sample starting with recording one signal on the YSZ substrate and subsequently probing the film in 5 nm steps in direction of the film surface. When the probe was closer to surface, the scanning step became 2nm steps for detailed study in surface structure. The single scans were shifted in the y direction for clarity. A typical series of line scans covering oxygen and cobalt signals is shown in Figure 6b and the acquisition positions are labeled in Figure 6a. From the spectra, it is shown that only the 6nm surface layer exhibits Co $L_{2,3}$ edge signals at 795 eV and 780 eV respectively. No significant Co signals at these two energy positions are observed in the deep area through the HfO_2 film from EELS analysis. This means that Co only segregates in the surface and forms a Co-rich surface layer, which is also confirmed by the x-ray photoemission spectroscopy (XPS) study. The fine structure of O-*K* edge in the HfO_2 film differs from the O-*K* edge in the YSZ revealing a change in the bonding structure. It should be noted that a small peak of O K-edges at 552 eV (denoted by an arrow) disappears in the 6nm surface layer where the Co $L_{2,3}$ edge appears. This implies that plenty of oxygen vacancies coexist with the possible substitution of Co atoms (2+ or 3+) to Hf (4+) atoms in the surface layer.

Room temperature magnetization experiment has been performed on this 40nm 5%Co:HfO_2 thin film and obtained good magnetic saturation value(M_s) [45]. We have found the surface enrichment of Co. This result makes a concept that the room temperature FM of this thin film most probably comes from the Co-rich oxide in a certain type of structure. This phenomenon is similar to the recent result on Co: TiO_2 reported by Chambers et al.in their films grown by oxygen plasma assisted MBE. However, in their experiment, the Co presents in the form of tiny surface clusters presumably of the CoTiO3 type. In our case the surface phase is still in the form of a uniform surface layer and high-resolution Z-contrast imaging observation throughout the film confirms no presence of clusters of the dopant. This result also implies the limited solubility of cobalt in HfO_2 under the processing conditions used. The room temperature FM reported under different levels of transition metallic dopants has to be carefully addressed.

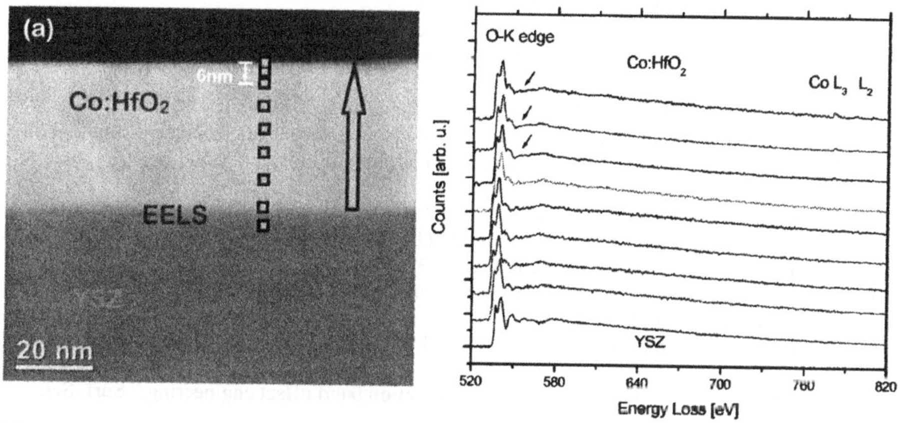

Figure 7 (a) A cross-sectional STEM Z-contrast image showing the high spatial resolution EELS acquisition area in the Co:HfO_2 thin film. (b) EEL spectra acquired across the interface in Co:HfO_2 thin film samples showing the distribution of Co element and others.

4. CONCLUSION

The microstructural characterization techniques of interface and defects with atomic scale analysis using the conjunction of STEM Z-contrast imaging and EELS have been well established on commercially available microscopes. In this paper, we have demonstrated the application of these advanced techniques into the structural and chemical analyses at the interface and defects in the newly developing transition-metal oxide heterointerfaces. The atomic and electronic structure can be determined through a combination of directly interpretable Z-contrast imaging and the corresponding EEL spectroscopy. Fast characterization and analysis based on the complex interfaces as well as defects in the advanced materials provide the basis for predictive material design with more desirable physical properties.

ACKNOWLEDGEMENTS

This research was sponsored by the National Science Foundation (NSF) under grant No. DMR-0335364. The microscopy work was performed at the National Center for Electron Microscopy, Lawrence Berkeley National Laboratory and supported by the Department of Energy (DOE) under Contract No. DE-AC02-05CH11231. Dr. T. Mizoguchi is gratefully acknowledged for helpful discussion in EELS analysis. We also thank the shared experimental facilities (SEFs) of pulsed laser deposition and Pelletron accelerator supported by the Center for Superconductivity Research and UMD NSF-MRSEC grant DMR 00-80008 for the sample fabrication.

REFERENCES

1. A. Ohtomo, H. Y. Hwang, "A high-mobility electron gas at the $LaAlO_3/SrTiO_3$ heterointerface," Nature, 427 (6973), 423-26 (2004).
2. R. F. Klie, J. P. Buban, M. Varela, A. Franceschetti, C. Jooss, Y. Zhu, N. D. Browning, S. T. Pantelides, S. J. Pennycook, "Enhanced current transport at grain boundaries in high-T-c superconductors," Nature, 435 (7041), 475-478 (2005).
3. S. R. Shinde, S. B. Ogale, J. S. Higgins, H. Zheng, A. J. Millis,V. N. Kulkarni, R. Ramesh, R. L. Greene, T. Venkatesan, "Co-occurrence of superparamagnetism and anomalous Hall effect in highly reduced cobalt-doped rutile TiO2-delta films, Phy. Rev. Lett., 92 (16), 166601 (2004).
4. R. F. Klie, Y. Zhu, E. I. Altman, Y. Liang, "Atomic structure of epitaxial $SrTiO_3$-GaAs(001) heterojunctions," App. Phys. Lett., 87 (14), 143106 (2005).
5. Z. Y. Chen, A. Biswas, I. Zutic, T. Wu, S. B. Ogale, R. L. Greene,T. Venkatesan, "Spin-polarized transport across a $La_{0.7}Sr_{0.3}MnO_3/YBa_2Cu_3O_{7-x}$ interface: Role of Andreev bound states," Phys. Rev. B, 63 (21), 212508 (2001).
6. G. P. Das, "Electronic-structure of epitaxial interfaces," Pramana – J. of Phys., 38, 545 (1992).
7. A. Franciosi, C. G. VandeWalle, "Heterojunction band offset engineering," Surf. Sci. Rept. 25, 1 (1996).
8. M. Peressi, N. Binggeli, and A. Baldereschi, "Band engineering at interfaces: theory and numerical experiments", J. of Phys. D – Appl. Phys., 31, 1273 (1998).
9. R. F. Klie, M. M. Disko, and N. D. Browning, " Atomic scale observation of the chemistry at the metal-oxide interface in heterogeneous catalysts," J. of Catal., 205, 1-6 (2002).
10. D. Kan, T. Takahito, Y. Shimakawa, M. Takano, "Fabrication and I-V characteristics of p-n junctions composed of high-Tc superconductors and La-doped $SrTiO_3$," Thin Solid Films, 486, 71-74 (2005).
11. S. D. Bu, B. H. Park, B. S. Kang, S. H. Kang, and T. W. Noh, .Influence of the Laser Fluence on the Electrical Properties of Pulse-Laser-Deposited $SrBi_2Ta_2O_9$ Thin Film, Appl. Phys. Lett. 75, 1155 (1999).

12. X. F. Du and I. W. Chen, "Fatigue of $Pb(Zr_{0.53}Ti_{0.47})O_3$ ferroelectric thin films," J. Appl. Phys. 83(12), 7789 (1998)

13. S. Iaiche S, N. Benouattas, A. Bouabellou, L. Osmani L, L. Salik, "Atomic diffusion at the Cu-Au-Si multilayers interface," Micron. Engr., 81 (2-4), 349-352 (2005).

14. Z. L. Zhang, W. Sigle, F. Phillipp, and M. Ruhle, "Direct atom-resolved imaging of oxides and their grain boundaries," Science 302, 846 (2003).

15. D.A. Muller, N. Nakagawa, A. Ohtomo, J.L. Grazul, and H.Y. Hwang, "Atomic-scale imaging of nanoengineered oxygen vacancy profiles in $SrTiO_3$," Nature 430, 657 (2004).

16. N. D. Browning, M. F. Chisholm and S. J. Pennycook, "Atomic-resolution chemical-analysis using a scanning-transmission electron-microscope," Nature 366, 143 (1993).

17. A. Ziegler, J. C. Idrobo, M. K. Cinibulk, C. Kisielowski, N. D. Browning, R. O. Ritchie, "Interface Structure and Atomic Bonding Characteristics in Silicon Nitride Ceramics", Science, 306, 1770 (2005).

18. I. Arslan I, T. J. V. Yates, N.D. Browning,P. A. Midgley, "Embedded nanostructures revealed in three dimensions," Science, 309, 2195-2198 (2005).

19. J. Bradley, Z. R. Dai, R. Erni, N. D. Browning, G. Graham, P. Weber, J. Smith, I. Hutcheon, H. Ishii, S. Bajt, C. Floss, F. Stadermann, S. Sandfords, "An astronomical 2175 angstrom feature in interplanetary dust particles," Science, 307 (5707), 244-247 (2005).

20. M. M. Mcgibbon, N.D. Browning, M. F. Chisholm, A. J. Mcgibbon, S. J. Pennycook, V. Ravikumar, V. P. Dravid, "Direct determination of grain-boundary atomic-structure in $SrTiO_3$," Science, 266 (5182), 102-104 (1994).

21. N. D. Browning, E. M. James, K. Kishida, I. Arslan, J. P. Buban, J. A. Zaborac, S. J. Pennycook, Y. Xin and G. Duscher, "Scanning Transmission Electron Microscopy: An Experimental Tool for Atomic Scale Interface Science," Reviews on Advanced Materials Science, 1, 1 (2000).

22. D. P. Norton, C. Park, C. Prouteau, D. K. Christen, M. F. Chisholm, J. D. Budai, S. J. Pennycook, A. Goyal, E. Y. Sun, D. F. Lee, D. M. Kroeger, E. Specht, M. Paranthaman, and N. D. Browning, "Epitaxial YBCO Films on Rolled-Textured Metals for High-Temperature Superconducting Applications," Mater. Sci. and Engr. B 56, 86 (1998).

23. C. Prouteau, G. Duscher, N. D. Browning and S. J. Pennycook, "Investigation of the Local Superconducting Properties in Ag-Sheathed BSCCO Tapes By STEM," Physica C 298, 1 (1998).

24. N. D. Browning, J. P. Buban, P. D. Nellist, D. P. Norton and S. J. Pennycook, "The Atomic Origins of Reduced Critical Currents at [001] Tilt Grain Boundaries in YBCO Thin Films," Physica C 294, 183 (1998).

25. R. F. Klie, Y. Ito, S. Stemmer and N. D. Browning, "Direct observation of oxygen vacancy ordering and segregation in perovskite oxides", Ultramicroscopy, 86, 289-302 (2001).

26. R. F. Klie and N. D. Browning, "Atomic scale characterization of a temperature dependence to oxygen vacancy segregation at $SrTiO_3$ grain boundaries", Appl. Phys. Letts 77, 3737-3739 (2000).

27. N. D. Browning, D. J. Wallis, P. D. Nellist and S. J. Pennycook, "EELS in the STEM: Determination of Materials Properties on the Atomic Scale," Micron 28, 333 (1997).

28. Y. Xin, E. M. James, I. Arslan, N. D. Browning, S. J. Pennycook, and S. Sivananthan, "Atomic Resolution Z-Contrast Imaging of Semiconductors," J. Electron Microscopy 49, 231 (2000).

29. D. Kan, T. Terashima, Y. Shimakawa, M. Takano, "Fabrication and I-V characteristics of p-n junctions composed of high-Tc superconductors and La-doped SrTiO3," Thin Solid Films, 486, 71 (2004).
30. Y. Muraoka, T. Muramatsu, J. Yamaura, and Z. Hiroi, " Photogenerated hole carrier injection to YBCO in an oxide heterostructure," Appl. Physc. Lett., 85, 2950 (2004).
31. R. Ramesh, D. M. Hwang, T. Venkatesan, T. S. Ravi, L. Nazar, A. Inam, X. D. Wu, B. Dutta, G. Thomas, A. F. Marshall, T. H. Gaballe, "Direct observation of structural defects in laser-deposited superconducting Y-Ba-Cu-O thin-films," Science 247, 57 (1990).
32. W. Ramadan,, S. B. Ogale1, S. Dhar, L. F. Fu, S. R. Shinde, Darshan C. Kundaliya, M. S. R. Rao, N. D. Browning, and T. Venkatesan, "Electrical properties of epitaxial junctions between Nb:SrTiO3 and optimally doped, underdoped and Zn-doped $YBa_2Cu_3O_{7-\delta}$," accepted by Phys. Rev. B, (2005).
33. W. Prellier, A. Fouchet, and B. Mercey, " Oxide-diluted magnetic semiconductors: a review of the experimental status," J. Phys.: Condens. Matter., 15, 1583 (2003).
34. H. Ohno, Science, "Making nonmagnetic semiconductors ferromagnetic," 281, 951 (1998).
35. F. C. Peiris, S. Lee, U. Bindley, J. K. Furdyna, "A prism coupler technique for characterizing thin film II-VI semiconductor systems," J. Appl. Phys., 84, 5194 (1998).
36. Y. Matsumoto, M. Murakami, T. Shono, T. Hasegawa, T. Fukumura, M. Kawasaki, P. Ahmet, T. Chikyow, S. Koshihara, H. Koinuma, "Room-temperature ferromagnetism in transparent transition metal-doped titanium dioxide," Science, 291 (5505), 854-856 (2001).
37. S. A. Chambers, S. Thevuthasan, R. Farrow, R. F. Marks, J. U. Thiele, L. Folks, M. G. Samant, A. J. Kellock, N. Ruzycki, D. L. Ederer, U. Diebold, "Epitaxial growth and properties of ferromagnetic co-doped TiO_2 anatase," Appl. Phys. Lett., 79 (21), 3467-3469 (2001).
38. J.K. Furdyna. "Diluted magnetic semiconductors," J. Appl. Phys. 64, R29 – R64 (1988).
39. A. Mauger and C. Godart, "The magnetic, optical, and transport properties of representatives of a class of magnetic semiconductors: The europium chalcogenides," Phys. Rep. 141, 51-176 (1986).
40. M.Venkatesan, C. B. Fitzgerald, J.M.D. Coey, "Unexpected magnetism in a dielectric oxide," Nature, 430, 630 (2004).
41. A. Oiwa et al. "Effect of spin injection on ferromagnetically coupled Mn spins in the III-V magnetic alloy semiconductor (Ga,Mn)As." Phys.Rev.Lett. 88, 1372021-024 (2002).
42. G.P.Das et al. "Ferromagnetism in Mn-doped GaN: From clusters to crystals," Phys.Rev.B. 68, 0352071-077 (2003).
43. T. Fukumura. "An oxide-diluted magnetic semiconductor: Mn doped ZnO." Appl.Phys.Lett. 75, 3366-3368 (1999).
44. K. Ando, "Magneto-optical properties of ZnO-based diluted magnetic semiconductors." J.Appl. Phys. 89, 7284-7286 (2001).
45. M. S. R. Rao, D. C. Kundaliya, S. Dhar, S. B. Oagle, S. R. Shinde, L. F. Fu, S. Welz, S. Lofland, N. D. Browning, and T. Venkatesan, "On the ferromagnetism in undoped and cobalt doped HfO_2". Appl. Phys. Lett. (to be submitted).

Author Index

Printed and bound by CPI Group (UK) Ltd, Croydon, CR0 4YY
16/09/2026
14963458-0003